# Highly Permeable Membranes

# Contributions to Nephrology

Vol. 46

Basel · München · Paris · London · New York · New Delhi · Singapore · Tokyo · Sydney

International Symposium on High-Efficiency Membranes for Optimised Therapy of Renal Failure, Bad Homburg v.d. Höhe, May 24–25, 1984

# Highly Permeable Membranes

Volume Editors
*E. Streicher*, Stuttgart
*G. Seyffart*, Bad Homburg v.d. Höhe

79 figures and 35 tables, 1985

 KARGER

Basel · München · Paris · London · New York · New Delhi · Singapore · Tokyo · Sydney

# Contributions to Nephrology

National Library of Medicine, Cataloging in Publication
International Symposium on High-Efficiency Membranes for Optimised Therapy of Renal Failure (1984: Bad Homburg von der Höhe, Germany)
Highly permeable membranes/
International Symposium on High-Efficiency Membranes for Optimised Therapy of Renal Failure, Bad Homburg v.d. Höhe, May 24–25, 1984;
volume editors, E. Streicher, G. Seyffart.
– Basel; New York: Karger, 1985. –
(Contributions to nephrology; v. 46)
Includes index.
1. Hemodialysis – congresses 2. Kidney Failure, Chronic – therapy – congresses 3. Membranes, Artificial – congresses I. Streicher, E. II. Seyffart, G. III. Title IV. Series
W1 CO778UN v. 46 [WJ 342 I595h 1984]
ISBN 3-8055-3994-0

Drug Dosage
The authors and the publisher have exerted every effort to ensure that drug selection and dosage set forth in this text are in accord with current recommendations and practice at the time of publication. However, in view of ongoing research, changes in government regulations, and the constant flow of information relating to drug therapy and drug reactions, the reader is urged to check the package insert for each drug for any change in indications and dosage and for added warnings and precautions. This is particularly important when the recommended agent is a new and/or infrequently employed drug.

# Contents

Contents

## Biocompatibility

## Hemodynamics

## Clinical Applications

# Foreword

After more than two decades of technical status quo in dialysis treatment, the development of highly permeable membranes and the introduction of filtration processes in the 1970s significantly improved the treatment strategies of end-stage renal failure.

Hemodialysis by diffusion using membranes made from regenerated cellulose has been supplemented by solute transport by means of convection (hemofiltration) and diffusion and convection combined (hemodiafiltration), thus improving the elimination of middle and large molecules. It is still assumed that these molecules contain uremic toxins despite certain controversy about the middle molecule theory. Experience has shown that these highly permeable, synthetic membranes are superior to conventional membranes, even in hemodialysis. In addition, reports have been published during the past few years on the improved biocompatibility of synthetic membranes.

In May 1984, an international symposium took place in Bad Homburg (FRG) in which 20 scientists discussed their experience in the application of highly permeable membranes in hemodialysis, hemofiltration and hemodiafiltration.

After a brief introduction describing the development and application of a new polysulfone membrane, lectures were given on the following topics: Characterisation of modern membranes (*Bosch* et al., *Stiller* et al., *Brunner* et al., *Sprenger* et al., *Wizemann* et al., *Röckel* et al.); Biocompatibility of modern membranes (*Aljama* et al., *Fawcett* et al., *Hildebrand* et al., *Stannat* et al., *Schaefer* et al., *Piazolo* et al); Changes in hemodynamics by modern membranes (*Schmidt* et al., *Schneider* et al., *Jahn* et al.) and Clinical application of modern membranes (*Fischbach* et al., *von Albertini* et al., *Klinkmann* et al., *Canaud* et al.). In these lectures and during the very lively discussions which followed, the advantages of synthetic membranes, which have been continually improved during the past few years, were stressed. In particular the results of tests on a new polysulfone membrane were considered to be extremely promising.

We should particularly like to thank Fresenius AG, Bad Homburg, FRG, who organised the symposium and provided generous financial support. We should also like to thank Karger-Verlag, Basel (Switzerland) for the high-quality printing of this book.

We hope that this symposium has contributed to further improvement in the treatment of patients with chronic renal insufficiency and has also drawn attention to new technologies and membrane materials.

August 1984

*E. Streicher*
*G. Seyffart*

# Introduction

Contr. Nephrol., vol. 46, pp. 1–13 (Karger, Basel 1985)

## The Development of a Polysulfone Membrane

A New Perspective in Dialysis?

*E. Streicher, H. Schneider*

Department of Nephrology and Hypertension, Katharinenhospital, Stuttgart, FRG

### Historical Review

Before presenting our results on the F 60 polysulfone membrane and the engineering aspects associated with the use of highly permeable membranes, I should like to give you some information on the history of the development of this membrane from our point of view.

The F 60 high-flux dialyser represents the temporary end of 10 years of development, testing and application of highly permeable synthetic membranes. Encouraged by the reports by *Henderson* et al. [1], *Quellhorst and Plashues* [2] and *Funck-Brentano* et al. [3] on the improvement in the elimination of large molecules achieved with the use of highly permeable membranes made out of polysulfone, cellulosenitrate and polyacrylonitrile, the membrane engineers of the Berghof Research Institute, Tübingen (FRG), Dr. *Strathmann* and Dr. *von Mylius* contacted us. The Berghof Institute had some previous experience in the production of asymmetric hollow fibre membranes for waste water purification and desalination of seawater. Together, we applied for a grant from the Federal Ministry for Research and Technology (Bundesministerium für Forschung und Technologie). After confirmation of the grant (FE Project MT 216), we started with the development work in 1974. At first we produced a hollow fibre out of 'Nomex', an aromatic polyamide (DFE 1).

This fibre (fig. 1) had a lumen of 600 µm, a wall thickness of 200 µm and a finger-like substructure [4]. Dr. *von Mylius* produced these fibres in a 'home-made' spinning plant and made them into hemofilters by potting them into old dialyser housings. Together with the Institute for Anatomy and Pathology of Domestic Animals, Hohenheim University, we per-

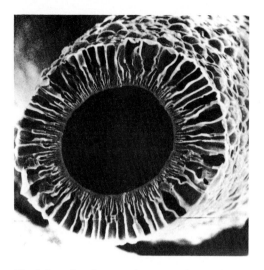

*Fig. 1.* Scanning electron microscope picture of the Berghof polyamide capillary (lumen 600 μm, wall thickness 250 μm). This asymmetric membrane shows a marked finger structure.

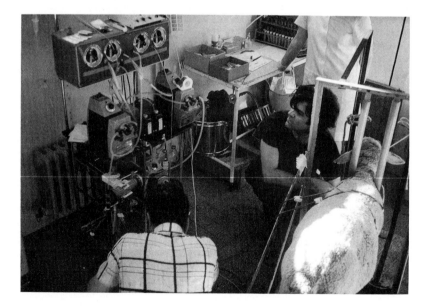

*Fig. 2.* The authors performing hemofiltration on a sheep using a polyamide capillary hemofilter of 0.4 m² surface area (Institute for Anatomy and Pathology of Domestic Animals, Hohenheim University).

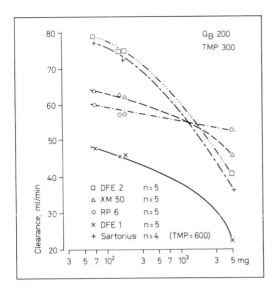

*Fig. 3.* Comparison of noncorrected in vivo plasma clearances of four different hemo-filters for urea, creatinine, uric acid and inulin. The performance increase observed with the Berghof polysulfone capillary membrane (upper curve) compared to the Berghof polyamide membrane (600 µm lumen) (lower curve) shows a marked gain in effectivity for all molecular ranges [6].

formed animal experiments with a new pilot series (surface area 0.4 m$^2$) [5]. 2 female sheep named Frieda and Emma had to suffer from our future-oriented mode of treatment (fig. 2). Each of the animals underwent periodic hemofiltration. Much to our amazement both animals survived this treatment unharmed.

The next stage was the production of a capillary hemofilter with a larger surface suitable for the treatment of ESRD patients. The DFE 1 resulted in a hemofilter with a surface of 1 m$^2$. The length of the capillary was 50 cm. A regular hemofiltration treatment with this filter was started in 3 patients in late 1975. The filtrate flow and the clearance rates for this product were still comparatively low (urea 67 ml/min, creatinine 65 ml/min, uric acid 67.5 ml/min, phosphate 62 ml/min, n = 17). To obtain more effective membranes the manufacture of a polysulfone hollow fibre was started in 1975, resulting in a reduction of the inner lumen to 250 µm. Changes in the geometry and membrane substance permitted much higher elimination rates [6]. Figure 3 compares the performance data of the DFE 2 polysulfone membrane with the DFE1, the old polyamide membrane.

During this initial phase we encountered mainly two problems. The manual procedure did not permit a very close potting of the fibres and we observed pyrogenic reactions with the filters produced under laboratory conditions. Therefore, in 1978 we contacted Messrs. Fresenius and asked them to process the polysulfone capillary produced by Berghof into a filter under professional conditions. At the same the fibre length was reduced to 32 cm, based on our optimising trials with hemofilters. The first results were presented by *Schneider* et al. [7] at the second ISAIO Congress in 1979.

### Present Device

Since the laboratory conditions did not permit a consistent standard-ised manufacture, variations in performance were frequently observed. Thus, in 1980, Mr. *Heilmann* and Mr. *Nederlof* of Fresenius (St. Wendel) started the development of a spinning plant with the aim to produce a uni-form quality under standardised and exact, reproducible production condi-tions. This development resulted in the F 60 polysulfone capillary which Fresenius provided us for initial clinical trials in 1982. The extent of the development realised by the St. Wendel engineers can only be understood when comparing the data of hemofiltration with the 1978 Berghof/Fre-senius co-production to the F 60 capillary (table I).

Scanning electron microsope studies of the fibres show the specific changes. The construction of the Berghof polysulfone fibre (fig. 4) still shows a marked finger-like structure and the formation of closed lagoons. Furthermore it is covered by a relatively compact inner and outer boundary layer. The F 60 membrane (fig. 5) is of uniform structure. The finger-like structure has been changed into a porous foam structure. At the lumen side the membrane (fig. 6) is evenly covered by pores of identical size. Without any boundary layer, the membrane turns directly from the substructure into the membrane surface which is interspersed with numerous pores (fig. 7).

Four features of the membrane structure contribute to the good trans-portation characteristics: the relatively low membrane thickness of 40 µm, the great porosity of the inner membrane layer, the regular porous foam structure of the membrane construction and the lack of a more dense outer boundary layer.

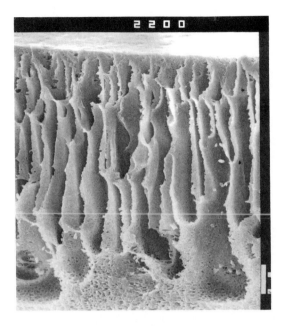

*Fig. 4.* Scanning electron microscope picture of the Berghof polysulfone membrane. The membrane has a marked finger-like structure. It shows a relatively compact inner and outer boundary. × 1,000.

*Table I.* Comparison between geometry and clearances in hemofiltration performed with the Berghof polysulfone capillary processed by Fresenius and the respective data for the Hemoflow F 60 at a blood flow of 300 ml/min

|  | DFE (Berghof-Fresenius) | Hemoflow F 60 |
|---|---|---|
| Surface | $0.9\,m^2$ | $1.25\,m^2$ |
| Lumen | 250 μm | 200 μm |
| Thickness | 200 μm | 40 μm |
| Clearance, ml/min |  |  |
| BUN | 73 | 119 |
| Creatinine | 78 | 118 |
| Phosphate | 76 | 117 |
| Inulin | 31 | 120 |
| Flux | 75 | 120 |

*Transport Characteristics*

The high clearance rates in filtration are partly due to the high hydraulic permeability of the membrane (fig. 8).

Already at a transmembrane of 300 mm Hg, the membrane is fully utilised. Further increases in pressure result in little gain only. At a TMP of 300 mm Hg at a blood flow of 100 ml/min we found an average filtration rate of 70 ml/min, at a blood flow of 200 ml/min a filtrate flow of 110 ml/min and at a blood flow of 300 ml/min a filtrate flow of 140 ml/min.

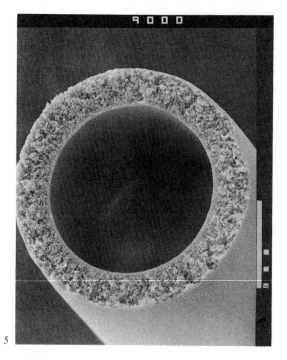

*Fig. 5.* Scanning electron microscope photograph of a cross section of the F 60 membrane. The membrane is of a uniform foam like structure. The inner and outer side merge into the foam structure without any compact boundary layer. × 300.

*Fig. 6.* Scanning electron microscope picture of the inner layer of the F 60 membrane. This separation layer is homogenously interspersed with pores of the same size. × 10,000.

*Fig. 7.* Scanning electron microscope picture of the outer surface of the F 60 membrane. × 10,000.

6

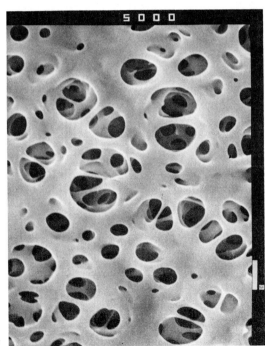

7

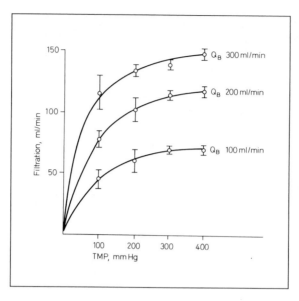

*Fig. 8.* In vivo ultrafiltration rates determined for the F 60 membrane at various transmembrane pressures and blood flows of 100, 200 and 300 ml/min, respectively.

The initial aim was the development of a hemofiltration membrane. However, the performance data of the F 60 in diffusive solute transport show that in dialysis it is also superior to all other membranes in the high molecular range.

We compared (table II) the in vivo plasma clearances for urea, creatinine, phosphate, inulin and $\beta_2$-microglobulin at a blood flow of 200 ml/min for hemofiltration with the clearances in dialysis at a dialysate flow of 500 ml/min, and hemodiafiltration at an additional filtrate flow of 50 ml/min. The gain in effectivity with hemodiafiltration was relatively low for small and large molecular substances, except for $\beta_2$-microglobulin where a sieving effect could already be demonstrated with a sieving coefficient of 0.79 [8]. We felt that the dialysis clearances for inulin at 85 ml/min and $\beta_2$-microglobulin at 56 ml/min were remarkably high. We investigated whether or not these findings which differed from the performance data of other membranes could be explained by the fact that in diffusive solute transport the F 60 polysulfone membrane does not interact at all or only insignific-

*Table II.* Mean plasma clearance values for urea, creatinine, phosphate and inulin and the sieving coefficient from 12 treatments each

| Mode | Urea | Cr | P | Inulin | $\beta_2$-Microglobulin |
|------|------|------|------|--------|------------------------|
| HD | 189 | 168 | 157 | 85 | 56 |
| HDF | 191 | 173 | 165 | 93 | 81 |
| HF | 119 | 118 | 117 | 120 | 95 |
| $S_c$ | 0.995 | 0.967 | 1.003 | 1.056 | 0.791 |

HD = hemodialysis: $Q_B$ 200, $Q_D$ 500, $Q_F$ 0 ml/min; HDF = hemodiafiltration: $Q_B$ 200, $Q_D$ 500, $Q_F$ 50 ml/min; HF = hemofiltration: $Q_B$ 200, $Q_F$ 120 ml/min. The sieving coefficients were determined under hemofiltration [8].

antly with the solutes transported, so that it closely resembles an ideal 'aqueous' membrane. We used a mathematical model developed by *Timmermann* et al. [9], Fraunhofer Institut für Grenzflächentechnik, Stuttgart, for the computer calculation of dialyser performance data.

We fed the specific geometric data of the F 60 capillary into the computer and calculated a dialysis clearance of 86 ml/min at a blood flow of 200 ml/min under the hypothetic assumption of a lacking membrane interaction for inulin. Based on these comparisons we feel that the F 60 polysulfone membrane meets the requirements for an ideal boundary layer. Further increases in the diffusive clearance can only be realised with additional reductions in the membrane thickness.

### Biocompatibility

At present, biocompatibility is receiving special attention in the evaluation of new membranes for dialysis und filtration. The behaviour of leukocytes and complement fractions is used as a standard. In sequential controls of leukocytes, thrombocytes, $C_3$ and $C_4$ complement during dialysis we found no significant changes (table III). The F 60 polysulfone membrane must be rated biocompatible according to the presently valid definition [12].

*Table III.* Values for leukocytes, thrombocytes, $C_3$ and $C_4$ complement at sequential measuring before the onset of the treatment, after 15, 60 and 180 min with the F 60 membrane

| n = 12 | Leucocytes | Thrombocytes $\times 10^3$ | $C_3$ mg/dl | $C_4$ mg/dl |
|--------|-----------|---------------------------|-------------|-------------|
| Before | 7,760 ± 1,503 | 173 ± 35 | 70,0 ± 5,8 | 68,7 ± 14,5 |
| 15 | 7,380 ± 1,434 | 164 ± 24 | 69,5 ± 5,6 | 67,7 ± 13,0 |
| 60 | 7,740 ± 1,297 | 159 ± 24 | 71,5 ± 3,4 | 70,1 ± 16,6 |
| 180 | 8,110 ± 2,169 | 169 ± 33 | 77,8 ± 6,3 | 73,3 ± 18,3 |

## New Engineering Aspects

In dialysis highly permeable membranes can only be used in closed dialysate systems. A pressure is maintained in the dialysate compartment which prevents a filtration. With an increasing hydraulic permeability of a membrane, the dialysate pressure must increasingly adapt to the pressure in the blood compartment. Since a pressure decrease takes place due to internal resistance from the arterial inlet to the venous outlet, the dialysate pressure at the venous outlet may increase above the pressure in the capillary. Basically this consideration applies to all dialysers. However, for highly permeable dialysers such as the F 60 membrane it is of special technical relevance. We registered the pressures during dialysis without additional filtration in the blood compartment before and after the F 60 capillary and the corresponding pressures in the dialysate compartment by means of continuous recording. As shown in figure 9, the blood pressure at the arterial inlet exceeded the dialysate pressure by 60 mm Hg, whereas at the end of the dialyser the dialysate pressure was 20 mm Hg higher than the corresponding pressure in the blood compartment. Under these conditions a filtration occurs at the beginning of the capillary and a backfiltration takes place at the end. Based on the pressure difference measured and the ultrafiltration factor of 48 ml/mm Hg/h determined by us at a blood flow of 200 ml/min for the F 60 capillary, we calculated an internal filtrate flow of 6 ml/min at a blood flow of 200 ml/min. This calculation does not consider the oncotic pressure of the blood. Here no exact values can be given without taking direct measurements at the membrane. This limitation shows clearly that the calculations of the internal filtration of highly perme-

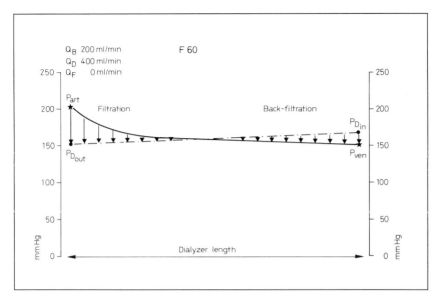

*Fig. 9.* Pressures measured simultaneously in hemodialysis without additional filtration in the blood compartment and the dialysate compartment at the arterial inlet and the venous outlet ($Q_B$ 200 ml/min, $Q_D$ 400 ml/min, $Q_F$ 0). At the arterial inlet the pressure in the blood compartment is higher than in the dialysate compartment, at the venous outlet the reverse condition is present. Under these conditions a filtration is taking place at the beginning of the capillary and a back-filtration at the end.

able membranes are primarily of a speculative nature. Here experimental studies of concentration dilutions of nonpermeating substances at defined points in the dialysate compartment should yield reliable results.

With an intact membrane, filtration and backfiltration are probably unproblematic. The severe problem seems to be that with a pressure in the blood compartment which is below the pressure in the dialysate compartment, unfiltered dialysate may enter into the blood in case of an undetected membrane leak. The above risk can be avoided when the dialysate pressure is reduced by filtration to such an extent that it is below the pressure of the blood compartment at the venous outlet. Several measurements showed (table IV) that at a blood flow of 200 ml/min and an hourly filtration rate of 500 ml/min the difference between dialysate pressure and venous outlet pressure will always become negative, generating conditions in which no dialysate may penetrate into the blood, not even in case of a leak.

*Table IV.* Dependence of pressure differences between dialysate pressure and pressure in the blood compartment on the hourly filtrate quantity with the F 60 membrane during dialysis

| Filtrate flux, ml/h | Pressure difference, $\Delta P$ (D-V) |
|---|---|
| 0 | $10.2 \pm 7.29$ |
| 100 | $9.8 \pm 7.08$ |
| 200 | $8.0 \pm 8.0$ |
| 300 | $4.4 \pm 6.98$ |
| 400 | $-0.8 \pm 3.34$ |
| 500 | $-2.8 \pm 2.68$ |
| 600 | $-7.6 \pm 5.72$ |
| 700 | $-11.6 \pm 6.69$ |
| 800 | $-13.6 \pm 9.52$ |

Relationship between filtrate flux rates and pressure difference (dialysate inflow minus blood outflow) using F 60 polysulfone capillary membrane. Blood flow 200 ml/min.

## Conclusions

Based on our results the following conclusion is drawn:

The performance data of the F 60 polysulfone membrane in hemofiltration, hemodiafiltration and hemodialysis have not yet been achieved by any other membrane. According to theoretical calculations, the transport characteristics of this membrane resemble an aqueous separating layer.

Considering kinetic aspects, filtration procedures are no longer superior to dialysis procedures when highly permeable membranes such as the F 60 are used. Measuring data of the transport characteristics however, presents only a partial aspect in the evaluation of the clinical effects. Final statements on the place value of a special procedure in chronic dialysis treatment will only be possible after several years. However, I hope this symposium will provide an answer to the question: 'Will the use of highly permeable dialysers present new clinical aspects in dialysis?'

## References

1    Henderson, L.W.; Besarab, A.; Michaels, A.; Bluemle, L.W.: Blood purification by ultrafiltration and fluid replacement (diafiltration). Trans. Am. Soc. artif. internal Organs *13:* 216 (1967).

2    Quellhorst, E.; Plashues, E.: Ultrafiltration: Elimination harnpflichtiger Substanzen mit Hilfe neuartiger Membranen; in: v. Dittrich, Skrabal, Aktuelle Probleme der Dialyseverfahren und der Niereninsuffizienz, vol. IV, p. 216 (Verlag Carl Bindernagel, Friedberg 1971).

3    Funck-Brentano, J.-L.; Sausse, A.; Man, N.K.; Granger, A.; Rondon-Nucete, M.; Zingraff, J.; Jungers, P.: Une nouvelle méthode d'hémodialyse associant une membrane à haute perméabilité pour les moyennes molécules et un bain de dialyse en circut fermé. Proc. Eur. Dial. Transplant Ass. 9: 55 (1972).

4    Mylius, U. v.; Streicher, E.; Schneider, H.W.: Kapillarmembranen zur Blutdiafiltration. Biomed. Tech. 21: 306 (1976).

5    Streicher, E.; Schneider, H.; Mylius, U. v.; Mahler, B.: Hämodiafiltration mit asymmetrischen Kapillarmembranen. Nieren- Hochdruck-Krankh. 5: 191 (1976).

6    Streicher, E.: Technik der Hämofiltration. Mitt. klin. Nephrol. VII: 55 (1978).

7    Schneider, H.; Streicher, E.; Mylius, U. v.: A theoretical and experimental approach towards optimal dimensions for capillary hemofilters. Proc. Artif. Organs 3: suppl., p. 114 (1979).

8    Streicher, E,; Schneider, H.: Polysulfone membrane mimicking human glomerular basement membrane. Lancet 11: 1136 (1983).

9    Timmermann, M.; Keller, H.-J.; Walitza, E.; Chmiel, H, H.: Numerische Simulation zum Hämodialyse- und Hämofiltrationsvorgang. Biomed. Tech. 26: 106 (1981).

10   Craddock, P.R.; Fehr, J.; Brigham, K.L.; Kronenberg, R.; Jacob, H.S.: Complement and leukocyte-mediated pulmonary dysfunction in hemodialysis. New Engl. J. Med. 296: 769 (1977).

11   Jacob, A.; Gavellas, G.; Zarco, R.; Perez, G.; Bourgoignie, J.: Leukopenia, hypoxia, and complement function with different hemodialysis membranes. Kidney int. 18: (1980).

12   Streicher, E.; Schneider, H.: Stofftransport bei Hämodiafiltration Nieren- Hochdruck-Krankh. 12: 339 (1983).

Dr. E. Streicher, Abt. f. Nieren- und Hochdruckkrankheiten,
Zentrum für Innere Medizin, Katharinenhospital, Kriegsbergstrasse 60,
D-7000 Stuttgart 1 (FRG)

# Membrane Characteristics

Contr. Nephrol., vol. 46, pp. 14–22 (Karger, Basel 1985)

# Effect of Protein Adsorption on Diffusive and Convective Transport Through Polysulfone Membranes

*T. Bosch, B. Schmidt, W. Samtleben, H.J. Gurland*

Department of Nephrology, First Medical Clinic, University Hospital
Munich-Grosshadern, Munich, FRG

*Introduction*

Diffusive transport is the major principle underlying blood purification by hemodialysis; it is especially effective in the elimination of substances in the low molecular weight range, e.g. of creatinine and urea. In contrast, hemofiltration is based on convective transport; this procedure exhibits excellent elimination characteristics especially for compounds of higher molecular weight. For some years now, research interest has focused on new synthetic membranes with the aim of combining the advantages of both techniques. This has led to the development of high flux membranes, in which high ultrafiltration rates are attainable and therefore the fraction of total transmembrane transport due to convection is increased as compared to conventional dialyzers.

Among other synthetic materials, polysulfone membranes have shown excellent permeability properties. *Colton* et al. [1], however, reported a marked permeability reduction of the polysulfone membrane by protein adsorption.

Therefore, it was the objective of this study to investigate the effect of protein adsorption on the polysulfone membrane in the Hemoflow F 60 (hemodiafilter, Fresenius, Bad Homburg, FRG, surface area A = 1.25 m², hollow fiber device) with special regard to changes both in ultrafiltration and clearances of creatinine and inulin.

## Materials and Methods

### UFR Measurements

Normal saline and plasma, respectively (37°C), were recirculated by a pump from a 3-liter cylinder through the blood compartment of the dialyzer at a flow rate of 200 ml/min. Prewarmed saline (37°C) was pumped from a plastic bag lying on electronic scales through the dialysate compartment ($Q_D$ = 500 ml/min). Pressures were monitored by Statham elements at the blood and dialysate in- and outlet ports. Protein concentration of plasma ranged from 4.4 to 4.9 g/dl. After each test period (10 min at any given TMP), the ultrafiltrate was substituted on the blood side in order to keep protein concentrations constant. Venous pressure and, concomitantly, TMP was varied between 0 and 100 mm Hg by a tube clamp. The UFR attained at a given TMP was determined gravimetrically and will be designated as 'net UFR' for reasons to be explained later. Net UFR was measured using normal saline, then pooled patient plasma followed again by normal saline. For comparison, experiments were carried out with the Asahi Hemofilter PAN 200 (A = 1.4 m$^2$); 3 dialyzers of each type were tested.

### Creatinine and Inulin Clearances

The blood side test solution (37°C) was recirculated from a 3-liter container standing on electronic scales through the dialyzer at a flow rate of 200 ml/min; dialysate was proportioned by a Centry II machine (Cobe, Lakewood, USA) and perfused the dialyzer at a flow rate of 500 ml/min in the single pass countercurrent mode. Each dialyzer was tested using blood side test solutions of normal saline followed by plasma. The initial creatinine and inulin concentrations of both the normal saline and plasma solutions were 100 and 50 mg/dl, respectively. Small quantities of [14]C-labelled creatinine and [3]H-labelled inulin were added to each test solution. Clearances were determined for both media at net UFRs between 10 and 50 ml/min according to equation 1 [2].

$$Cl = \frac{(Q_{Bin} - Q_F)(C_{Bin} - C_{Bout})}{C_{Bin}} + Q_F , \qquad (1)$$

where Cl = clearance, ml/min, $Q_{Bin}$ = blood flow rate at dialyzer inlet port, ml/min, $Q_F$ = ultrafiltration rate, ml/min, and $C_{Bin(out)}$ = blood concentration of solute at dialyzer inlet (outlet) port, mg/dl.

Inulin and creatinine concentrations were determined in samples drawn from the blood compartment before and after the dialyzer, by assessment of radiolabel activity with a liquid scintillation spectrometer (Betaszint BF 5000).

## Results

Figure 1 shows a plot of net UFR vs. TMP using the polysulfone membrane. The 3 regression lines shown were determined for each of the solutions tested in sequence; each line represents the pooled results for 3 dialyzers. As expected, very high UFRs of up to 80 ml/min were attained with

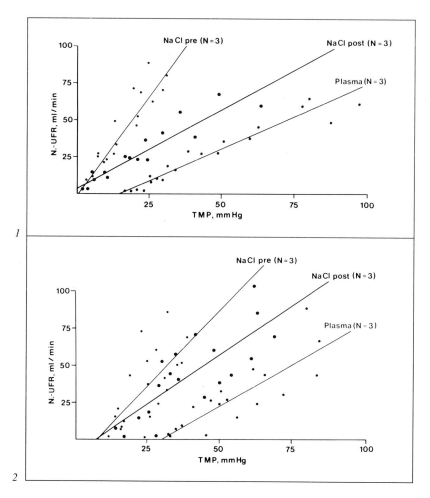

*Fig. 1.* Net UFR before, during and after plasma perfusion: Fresenius Hemoflow F 60 (A = 1.25 m²).

*Fig. 2.* Net UFR before, during and after plasma perfusion: Asahi Hemofilter PAN 200 (A = 1.4 m²).

normal saline. The use of plasma led to a marked decrease in the slope of the regression line, indicating a lower permeability; moreover, it is shifted to higher TMPs due to the oncotic pressure of the plasma proteins. The middle line in figure 1 represents the results of the last saline perfusion; its slope is comparable to that of the plasma line. There is a linear relation

*Table I.* Hydraulic permeability coefficients $K_p$, ml/min $\cdot$ mm Hg $\cdot$ m$^2$ of Hemoflow F 60 and Hemofilter PAN 200

| Device | Test Solvent | | |
|---|---|---|---|
| | $NaCl_{pre}$ | plasma | $NaCl_{post}$ |
| Hemoflow F 60 (1.25 m$^2$) | 2.10 | 0.68 | 0.85 |
| Hemofilter PAN 200 (1.4 m$^2$) | 1.45 | 0.81 | 0.94 |

between UFR and TMP (varied between 0 and 100 mm Hg) for all 3 experimental conditions, correlation coefficients exceeded +0.9.

Figure 2 shows the results of net UFR measurements in the Asahi PAN membrane which were analogous to those of the F 60. In comparison to results for the F 60, the zero UFR TMP intercept value for all lines is shifted towards higher TMPs. This is caused by flow obstacles incorporated in the Asahi Hemofilter housing at the dialysate in- and outlet ports which significantly increase the recorded TMP if pressure readings are taken outside the housing. However, this phenomenon has no influence on the interpretation of our results which are assessed on the basis of the slope of each regression line. Net UFR correlated with TMP for all 3 experiments by linear regression (r = 0.7).

Table I gives hydraulic permeability coefficients $K_p$ in ml/min $\times$ mm Hg $\times$ m$^2$ for both membranes before, during and after plasma perfusion. $K_p$ is defined as quotient of net UFR divided by TMP and surface area. The 'new' polysulfone membrane yields a markedly higher $K_p$ than the Asahi PAN membrane. Using plasma, however, the permeability of the polysulfone membrane dropped below that of the PAN membrane; during the final perfusion with saline the permeability of both membranes increased slightly.

Table II describes creatinine and inulin clearances using electrolyte solution and heparinized patient plasma respectively at net UFR = 0 and 50 ml/min. Without any net UFR, creatinine clearance averaged 173 ml/min. At a net UFR of 50 ml/min, creatinine clearance was increased by 6% to 183 ml/min. These results were found both with electrolyte solution and plasma as test medium.

Inulin clearance without net UFR amounted to 96 ml/min using electrolyte solution and 91 ml/min using plasma. At a net UFR of 50 ml/min, in-

*Table II.* Effect of ultrafiltration on creatinine and inulin clearance; hemodiafilter: Hemoflow F 60 (n = 3)

| | Creatinine clearance, ml/min | | Inulin clearance, ml/min | |
|---|---|---|---|---|
| | electrolyte solution | plasma | electrolyte solution | plasma |
| UFR = 0 ml/min | 173.3 ± 2.5 | 172.2 ± 2.8 | 95.6 ± 4.1 | 91.3 ± 8.4 |
| UFR = 50 ml/min | 183.4 ± 2.7 | 183.4 ± 4.0 | 117.2 ± 4.9 | 118.9 ± 3.4 |
| % clearance increment | + 5.8 | + 6.5 | + 22.6 | + 30.2 |

$Q_B$ = 200 ml/min; $Q_D$ = 500 ml/min.

ulin clearance in both media rose to 117 ml/min representing a 22 and 30% increase for electrolyte solution and plasma, respectively.

Figure 3 depicts schematically the pressure profiles along the polysulfone membrane using electrolyte solution as the 'blood' compartment solvent. Blood pressure drops rapidly along the length of the dialyzer membrane and eventually falls below the dialysate pressure. Therefore, in the second half of the dialyzer membrane, backfiltration of dialysate into the blood compartment takes place.

Results of a typical experiment are given in figure 4: using electrolyte solution in the blood compartment, blood and dialysate pressures were measured at the respective in- and outlet ports ($Q_B$ = 200, $Q_D$ = 500 ml/ min). Net UFR was adjusted between 10 and 50 ml/min. The overlapping area indicates that dialysate inlet pressure $P_{Di}$ exceeds blood outlet pressure $P_{Bo}$ and, therefore, backfiltration takes place. Higher TMP and higher UFRs lead to a decrease in backfiltration.

Figure 5 shows an analogous experiment carried out with plasma. In this case, at UFRs > 10 ml/min, dialysate inlet pressure is less than blood outlet pressure and thus no backfiltration takes place.

Figure 6 shows theoretical backfiltration rates ($UFR_2$) for the F 60 dialyzer plotted as a function of $\Delta_1 = P_{Bi} - P_{Do}$ and $\Delta_2 = P_{Di} - P_{Bo}$ which were calculated from equation 2.

$$UFR_2 = \frac{\Delta_2^2}{2(\Delta_1 + \Delta_2)} \cdot K_p \cdot A \qquad (2)$$

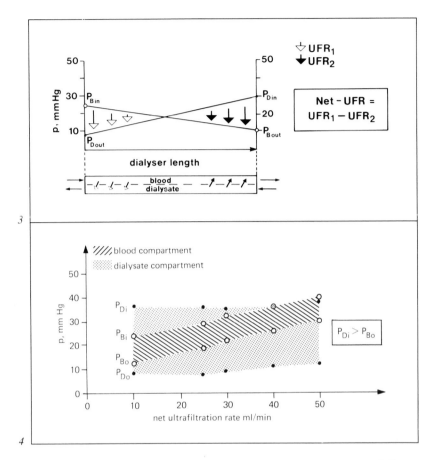

*Fig. 3.* Schematic pressure profiles in a high-flux polysulfone membrane hemodiafilter.

*Fig. 4.* Hemoflow F 60 in- and outlet pressures of blood and dialysate compartments as a function of net UFR using electrolyte solution.

where $UFR_2$ = backfiltration rate, ml/min; $\Delta_1 = P_{Bi} - P_{Do}$, mm Hg; $\Delta_2 = P_{Di} - P_{Bo}$, mm Hg; $K_p$ = hydraulic permeability coefficient, ml/min · mm Hg · m²; and A = surface area, m².

As an approximation, crossing pressure profiles were assumed to be linear and oncotic pressure was disregarded. The dotted square represents the scope of $\Delta_1$ and $\Delta_2$ determined from our experiments with theoretical $UFR_2$s amounting to maximum values of 10 ml/min using electrolyte solution and 5 ml/min using plasma.

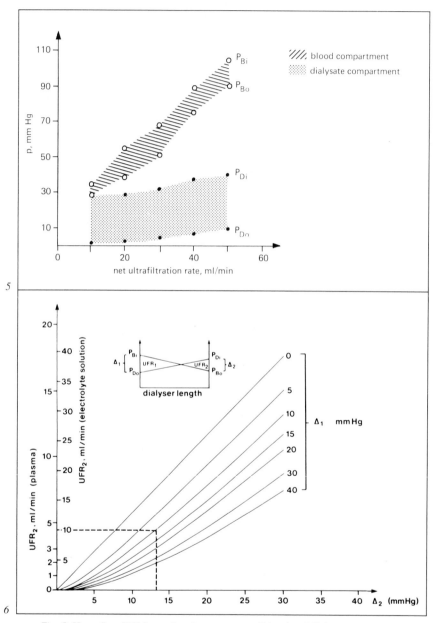

*Fig. 5.* Hemoflow F 60 in- and outlet pressures of blood and dialysate compartments as a function of net UFR using plasma.

*Fig. 6.* Hemoflow F 60: calculated backfiltration $UFR_2$ as a function of $\Delta_1$ and $\Delta_2$.

*Discussion*

Extremely high ultrafiltration rates of up to 80 ml/min were found (TMP $\leq$ 100 mm Hg) with both the F 60 polysulfone membrane and the Asahi PAN 200 using saline as solvent. Plasma perfusion caused an irreversible protein adsorption on the interior surface of the filter capillaries which led to a reduced permeability of the membrane. Further perfusion of the filter with saline could not desorb the protein layer from the membrane effectively as is demonstrated by the virtually unchanged $K_p$. Even though protein adsorption caused a 60% decrease of $K_p$ in the F 60 device, the absolute value of 0.7 ml/min $\times$ mm Hg $\times$ m$^2$ after plasma contact is comparable to $K_p$ in the Asahi Hemofilter (0.8 ml/min $\times$ mm Hg $\times$ m$^2$); at any rate, both $K_p$s exceed the hydraulic permeability coefficient of conventional dialyzers by one order of magnitude.

The relationship between net UFR and TMP was found to be linear not only for normal saline as test solvent but also for plasma. This means that within the applied TMP range between 0 and 100 mm Hg, the total mass transfer resistance for both membranes is determined by the transport resistance of the membrane itself. High UFR normally leads to a marked plasma resistance which is caused by the building-up of a protein boundary layer. This phenomenon, however, plays only a minor role at TMPs below 100 mm Hg. Moreover, plasma protein concentration was held reasonably constant by the substitution of the ultrafiltration volume after each 10-min test period.

Protein adsorption had no influence on creatinine elimination through the membrane, as is reflected by identical creatinine clearances using both saline and plasma. Inulin clearance by purely diffusive transport (UFR = 0) was found to be slightly lower using plasma (91 ml/min) as compared to saline (96 ml/min); however, at UFR = 50 ml/min, identical clearances of 118 ml/min were found for both media.

As expected, creatinine (MW 113) with lower molecular weight yields a higher clearance than inulin (MW 5,200); moreover, it is a generally known fact that clearance increment by additional convective transport (UFR > 0) is a function of molecular weight. Thus, 50 ml/min UFR increased inulin clearance by 22–30%, whereas creatinine clearance was raised by only 6%.

Inulin and creatinine clearances even at net UFR = 0 ml/min were extraordinarily high; this shed some doubt on the assumption of purely diffusive transport. Using electrolyte solution, a 'crossing-over' of pressure profiles was found (fig. 3, 4). Apparently, the very high UFR leads to a marked pressure drop on the 'blood side' of the membrane which can even fall below the dialysate pressure. This in turn causes a backfiltration of dialysate into the blood compartment of the dialyzer. With the resulting 'crossing-over' pressure profiles, the gravimetrically determined net UFR represents the difference of ultrafiltration into the dialysate compartment ($UFR_1$) minus backfiltration to the blood side of the membrane ($UFR_2$). This phenomenon is not seen in conventional dialyzers with moderate permeability where pressure in the blood compartment exceeds dialysate pressure at any point of the membrane; therefore, backfiltration is not possible.

In summary, even at net UFR = 0, there is some 'hidden' additional convective transport of blood solutes across the membrane which increases purely diffusive clearances.

Using plasma as test medium (fig. 5), no backfiltration takes place at UFR $\geq 10$ ml/min. This is due to the fact that the oncotic pressure of plasma proteins and the increased solvent viscosity (as compared to electrolyte solution) both raise the pressure on the blood side of the membrane ($Q_B = 200$ ml/min). Thus, $P_B$ remains greater than $P_D$ at any point of the membrane.

According to figure 6, theoretical backfiltration rates and, therefore, additional convective transport volumes were calculated to be $\leq 10$ ml/min using electrolyte solution and $\leq 5$ ml/min using plasma. In conclusion, the extremely high clearances achieved using the polysulfone membrane are not only due to increased convective transport, but also reflect its excellent diffusive transport characteristics.


*References*

1    Colton, C.K.; Henderson, L.W.; Ford, C.A.; Lysaght, M.J.: Kinetics of hemodiafiltra-
     tion. I. In vitro transport characteristics of a hollow fiber blood ultrafilter. J. Lab. clin.
     Med. *85:* 355–371 (1975).
2    Klein, E.; Autian, J.; Bower, J.D.; Buffaloe, G.; Centella, L.J.; Colton, C.K.; Darby,
     T.D.; Farrell, P.C.; Holland, F.F.; Kennedy, R.S.; Lipps, B.; Mason, R.; Nolph, K.D.;
     Villarroel, F.; Wathen, R.L.: Evaluation of hemodialyzers and dialysis membranes.
     DHEW Publ. No. (NIH) 77–1294, p. 46 (US Government Printing Office, Washington
     1977).

Dr. med. Dr. rer. nat. Thomas Bosch, Nephrologische Abteilung,
Medizinische Klinik I, Klinikum Grosshadern der Universität München,
Postfach 701260, D-8000 München 70 (FRG)

Contr. Nephrol., vol. 46, pp. 23–32 (Karger, Basel 1985)

# Backfiltration in Hemodialysis with Highly Permeable Membranes

*S. Stiller, H. Mann, H. Brunner*

Department of Internal Medicine II, Technical University of Aachen, FRG

## Introduction

Modern dialysis monitors control the pressure in the dialysis fluid in such a way that a fixed ultrafiltration rate is obtained. Control of ultrafiltration allows the use of highly permeable membranes for hemodialysis. When using dialysers with highly permeable membranes, the transmembrane pressures are very low and along the flow path a pressure, which is higher on the dialysate side of the dialyser than on the blood side, called a "negative" pressure, may occur. In this case, dialysis fluid is filtered back into the blood of the patient and the question arises wether backfiltration can cause potentially unsterile dialysis fluid or pyrogens to enter the blood of the patient.

In this paper the distribution of pressure and filtration rate along the flow path in a capillary dialyser is analyzed and for the F 60 hemodiafilter the calculated results are presented. From this knowledge, it can be decided if backfiltration may endanger the patient. A mathematical model was employed to verify the practical relevance of back filtration.

## The Mathematical Model

Figure 1 shows schematically the pressure on the blood and dialysate side along the flow path as a function of the distance (x) from the blood inlet (see table I). Blood enters at $x = 0$ with the pressure Pb(0) and due to the flow resistance in the capillaries drops to a pressure (Pv) at $x = L$, given by the flow resistance back in the patient. The oncotic pressure of the

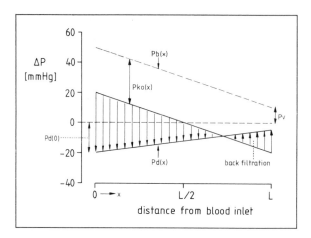

*Fig. 1.* Pressure gradient as a function of the distance (x) from blood inlet (schematically). The hydrostatic pressure inside the capillaries Pb(x) decreases with (x) due to the flow resistance to the pressure Pv at the blood outlet. The pressure in the dialysate compartment Pd(x) increases with (x), because dialyzing fluid enters at x = L and flows in the direction of the negative x-axis, Pd(0) is the dialysate pressure controlled by the dialysis machine. The sum Pb(x) + Pd(x) is the hydrostatic pressure gradient; to obtain the total pressure gradient, the onkotic pressure must be subracted: Pb(x) + Pd(x) − Pko(x).

plasma proteins Pko(x) subtracts from the hydrostatic pressure in blood. Dialysis fluid enters at x = L and flows in the negative direction of the x-axis. Due to the flow resistance in the dialysate compartment the pressure rises with (x). At the dialysate outlet the pressure equals the pressure Pd(0), generated by the monitor. The pressure gradient is given by the sum of the partial pressures Pb(x) − Pko(x) + Pd(x), indicated by the arrows in figure 1. If they point upward, pressure gradient is negative and backfiltration takes place.

The blood flow Qb(x) and the hydrostatic pressure in the blood compartment Pb(x) are described by two simple differential equations:

$$\frac{dQb\,(x)}{dx} = -\,q(x) \tag{1}$$

$$\frac{dPb\,(x)}{dx} = -\,w(x)\,Qb(x) \tag{2}$$

Blood flow decreases according to the local filtration rate q(x), hydrostatic pressure decreases proportional to the blood flow. The flow resistance w(x) is a function of the flow geometry and blood viscosity (see below). Local filtration rate q(x) is given by the product of the hydraulic permeability of the membrane (Lp) and the pressure gradient:

$$q(x) = Lp\,[Phy(x) - Pko(x)] \tag{3}$$

*Table I.* Symbols

| | |
|---|---|
| Cpr(x) | protein concentration |
| Cpy | concentration of pyrogens |
| Hk(x) | hematocrit |
| N | number of capillaries in the dialyser |
| Lp | hydraulic permeability filtration |
| Lpb | hydraulic permeability backfiltration |
| Pb(x) | hydrostatic pressure inside the capillary |
| Pd(x) | hydrostatic pressure in the dialyzing fluid |
| Phy(x) | hydrostatic pressure gradient between plasma and dialyzing fluid |
| Pko(x) | oncotic pressure inside the capillaries |
| Qb(x) | blood flow rate |
| Qd | dialysate flow rate |
| Ql | leakage rate |
| Qr | rate of backfiltration |
| S | sieving coefficient of pyrogens |
| u(x) | blood viscosity |
| w(x) | flow resistance |
| a1 | = 0.28 constant |
| a2 | = 0.0018 constant |
| a3 | = 0.000012 constant |
| b | pressure drop in the dialysate compartment<br>= 40 mm Hg/l/min |
| b0 | = 1.2 constant |
| b1 | = 3.175 constant |
| b2 | = −1.8478 constant |
| b3 | = 15.208 constant |
| 1 | effective length of the capillaries |
| r | inner radius of the capillaries |

The hydrostatic pressure difference Phy(x) is the sum of Pb(x) and Pd(x). It is assumed that Pd(x) increases linearly with (x). The hydraulic permeability of the membrane for the filtration (Lp) and the backfiltration (Lpb) are unequal because during filtration the plasma proteins reduce the membrane pore area. (Lp) may therefore change with time and pressure, but here is assumed to be constant.

$$Phy(x) = Pb(x) + Pd(x) \qquad (4)$$

$$Pd(x) = Pd(O) - b\, Qd\, x \qquad (5)$$

Oncotic pressure is caluclated as a third order polynominal from protein concentration [2]:

$$Pko(x) = a1\, Cpr(x) + a2\, Cpr(x)^2 + a3\, Cpr\, (x)^3 \qquad (6)$$

Since the plasma proteins cannot pass the membrane their concentration is a function of (x). The same is true for the hematocrit:

Flow in the capillaries occurs with a low Reynolds number of about 3 and therefore is laminar. Flow resistance w(x) is then given by the equation of Hagen und Poiseuille [6]:

$$w(x) = \frac{8\,u(Hk)}{3.14\,N\,r^{**}4} \qquad (9)$$

In equation 9 (N) is the number of the capillaries, (r) their internal radius, and u(Hk) is the blood vicosity. The function u(Hk) is introduced as a third order polynominal:

$$u(Hk) = b0 + b1\,(Hk(x) - 0.1) + b2\,(Hk(x) - 0.1)^2 \qquad (10)$$
$$+ b3\,(Hk(x) - 0.1)^3$$

The constants b0, b1, b2, b3 were calulated from the relation u(Hk) as given in *Diem and Lentner* [3]. For small radii blood viscosity is reduced by the Farae-Lundquist effect [5]:

$$u(r) = u(R)/(1 + d/r)^2 \qquad (11)$$

In equation 11 u(R) is the viscosity measurered in a tube with a radius $R \gg r$, and (d) is the diameter of the erythrocytes.

Solution of equations 1 and 2 is performed numerically by a Runge-Kutta algorythm. An appropriate program for a microcomputer has been written in Fortran.

For the hemodiafilter F 60 pressure and filtration along the flow path were calculated for different conditions in hemodialysis. Effective membrane area is 1.25 m², internal radius of the capillaries is 0.01 cm, their length is 23 cm, the number of capillaries is 8646. Ultrafiltration coefficient is as high as 40 ml/h/mm Hg, an increases for backfiltration by a factor of 2 to 2.5. A factor of 2 is used here.

## Results

Figure 2 shows the calculated hydrostatic pressure difference Phy(x), onkotic pressure Pko(x), and local filtration rate q(x). The parameters were chosen (blood flow Qb(0) = 200 ml/min, protein concentration Cpr(0) = 50 g/l, hematocrit Hk(0) = 0.45) to obtain backfiltration both by oncotic and negative hydrostatic pressure. Backfiltration starts at x = a, because the hydrostatic pressure difference Phy(x) decreases below the oncotic pressure Pko(x). For values x > b (hatched area) hydrostatic pressure difference is negative. Filtration in this example is 8.35 ml/min, backfiltration 6.75 ml/min, and ultrafiltration 1.61 ml/min.

Note that under unfavorable circumstances a negative hydrostatic pressure difference at the membrane indeed can occur. The quantity of po-

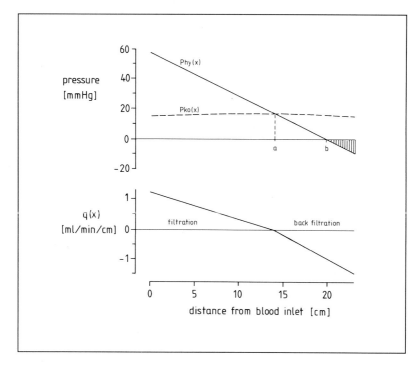

*Fig. 2.* Local filtration rate q(x), hydrostatic pressure gradient Phy(x), and oncotic pressure Pko(x) as function of the distance (x) from blood inlet. The calculation is based on the following data: Dialyser effective membrane area: 1.25 m²; capillary length: 23 cm; inner radius: 0.01 cm; blood flow rate: 200 ml/min; dialysate flow rate: 500 ml/min; hematocrit: 0.45; plasma protein concentration: 50 g/l; ultrafiltration coefficient for filtration: 40 ml/h/mm Hg; for backfiltration: 80 ml/h/mm Hg. The rate of filtration is 8.35 ml/min, of backfiltration is 6.75 ml/min, and ultrafiltration is 1.61 ml/min. Hydrostatic pressure gradient is negative in the hatched area.

tentially unsterile dialysing fluid which enters the bloodstream can be estimated. Let us assume that a leakage in a single capillary occurs at x = L with a negative pressure (P1) (fig. 3). Then the pressure in the blood at the site of the leakage is increased by (P1) to Pd(L). The blood flow in the defective capillary is reduced because the pressure drop from blood inlet to x = L is reduced by (P1). From the site of the leakage a mixture of blood and dialysis fluid (Qn + Ql) flows to the blood outlet. The pressure at blood outlet will not change if the dialysate flow into the blood is small compared to the blood flow. The pressure drop in the residual flow path from the leak

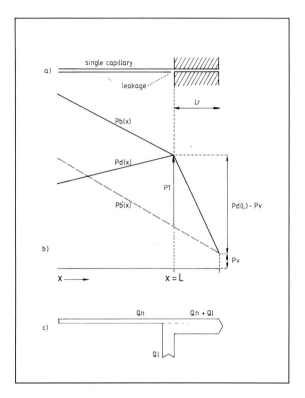

*Fig. 3.* Leakage in a single capillary. The pressure inside the capillary at the site of the leakage rises to the pressure in the dialysate Pd(L). Dialysate enters the bloodstream at a rate (QL) and the pressure drop (Pd(L) − Pv) causes a mixture of blood and dialysis fluid (Qn + Ql) flowing through the residual flow path (Lr).

to the blood outlet (Lr) is (Pd(L) − Pv). Thus the flow rate in the residual ·flow path (Ql + Qn) is determined by the pressure drop (Pd(L) − Pv) and the flow resistance. The rate at which dialysis fluid enters the blood can be estimated by equation 12:

$$Ql = k \ \frac{3.1248 \, P1 \, r^4}{8 \, u} \ \left(\frac{1}{Lr} + \frac{1}{L}\right) - \frac{Qb(L)}{N} \ .$$

In the derivation of equation 12 (Pd(L) − Pv) was replaced by P1. The viscosity (u) should be that of the mixture of blood and dialysis fluid, but for simplicity the viscosity of blood is used. In reality dialysis fluid will enter

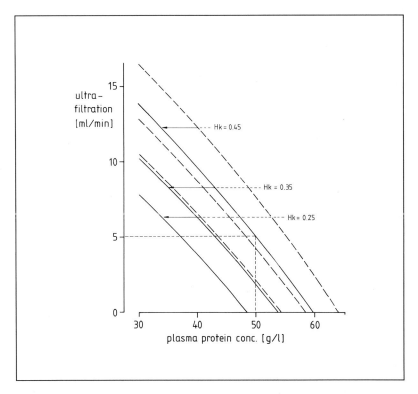

*Fig. 4.* Ultrafiltration rate which is necessary to avoid a negative hydrostatic pressure as a function of the plasma protein contration for hematocrit values between 0.25 and 0.45 and to dialysate loss rate (solid lines: 500 ml/min; dashes lines: 700 ml/min).

the blood at a somewhat higher rate than given by equation 12. The pressure P1 can reach values of 10 mm Hg (fig. 2). With this value and (r = 0.01 cm, u = 0.0235 Poise, Lr = 1 cm, L = 23 cm, Qb(L) = 200 ml/min, N = 8646, k = 79993.2) Ql is calculated as:

Ql = 0.116 ml/min

The conditions for a negative hydrostatic pressure to occur are calculated in figure 4. Ultrafiltration (on the ordinate) is calculated as a function of plasma protein concentration (abscissa) for three values of hematocrit (0.25, 0.35, 0.45), and for two values of dialysate flow rate (500 ml/min,

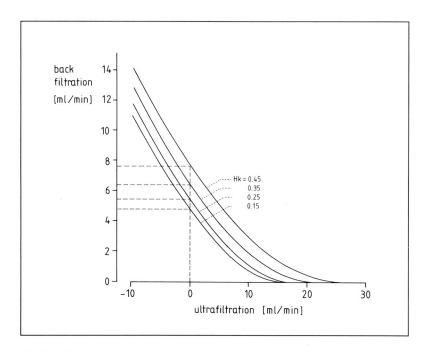

*Fig. 5.* Relation between backfiltration and net ultrafiltration for hematocrit values between 0.15 and 0.45. Plasma protein concentration is 66 g/l, the data are as in figure 2. To
suppress backfiltration, an ultrafiltration rate of 15.8 ml/min (948 ml/h) is necessary if the
hematocrit is 0.25.

solid lines and 700 ml/min, dashed lines). A negative hydrostatic pressure
occurs with low protein concentration, high values of hematocrit, and high
dialysate flow rate. With a plasma protein concentration 50 g/l and a
hematocrit of 0.45 an ultrafiltration rate of 5 ml/min is sufficient to avoid
a negative hydrostatic pressure. From figure 4 follows that with moderate
ultrafiltration rates the potentially dangerous negative hydrostatic pressure
can be avoided.

The amount of total backfiltration for a given ultrafiltration rate and
values of the hematocrit between 0.15 and 0.45 is calculated in figure 5
(blood flow 200 ml/min, protein concentration 66 g/l). Backfiltration for
zero ultrafiltration occurs at a rate between 4.5 and 7.5 ml/min, depending
on hematocrit. To avoid back filtration at all, ultrafiltration rates between
900 and 1,560 ml/h are necessary.

Looking at the influence of the capillary geometry on backfiltration, for a constant effective membrane area (blood flow 200 ml/min, protein concentration 66 g/l) the ultrafiltration rate to avoid backfiltration at all is calculated as a function of the capillary length for different values of the internal radius (0.01, 0.008, 0.006, and 0.005 cm) and two values of the hematocrit 0.25 and 0.35). Reduction in radius very rapidly increases the amount of back filtration. The reduction of blood viscosity with decreasing radius by the Faraeus-Lundquist effect [5] does not outweight the increase of the influence of the radius on flow resistance in the capillary.

*Discussion*

The mathematic analysis shows that backfiltration is nearly always present in hemodialysis with highly permeable membranes. Considering possible effects of backfiltration one has to distinguish between backfiltration caused by negative hydrostatic or a oncotic pressure. At a membrane leakage, the oncotic pressure is not effective, because at the side of the leakage the proteins can pass the membrane and therefore no pressure gradient can develop. If a positive hydrostatic pressure gradient exists at the leakage, erythrocytes will get into the dialysate and cause blood leak alarm. But if a leakage occurs at a site with a negative hydrostatic pressure gradient, dialysate will enter the blood and the leakage may not be detected.

Leakage rate per capillary can be more than 0.1 ml/min, which is several times the blood flow in one capillary (0.023 ml/min at Qb = 200 ml/min), and more than 1 capillary may be defective. Thus, an appreciable amount of potentially unsterile dialysate in principle could enter the blood. For this reason ultrafiltration rates should be kept high enough so that negative hydrostatic pressure gradients are avoided.

Since under normal conditions backfiltration by negative oncotic pressure is very common in hemodialysis with highly permeable membranes, the question should be answered if the transport of pyrogens from dialysis fluid into the blood may be enhanced. If pyrogens at a certain concentration (Cpy) are present in the dialysis fluid, the flux of pyrogens to the blood due to backfiltration is given by (S Cpy Qr), where (S) is the sieving coefficient, and (Qr) the backfiltration rate. The sieving coefficient for the F 60 dialyser is estimated to be S = 0.001 [4]. Backfiltration with rates not higher than 7.5 ml/min, therefore, can only transport minimal quantities to the blood. Pyrogenic substances will also enter the blood by diffusion, and the significance of the backfiltration for the transport of pyrogenes has to be considered in comparison with diffusive transport, but no permeability data for pyrogens exist for the F 60 membrane. A final conclusion with respect to pyrogens is therefore not yet possible.

## Conclusion

Backfiltration of dialysate occurs in hemodialysis with highy permeable membranes. Backfiltration must be considered to be potentially dangerous if it is caused by a negative hydrostatic pressure gradient, because then unsterile dialysing fluid could invade the bloodstream undetected. A negative pressure gradient is favored by a low plasma protein concentration and a high hematocrit. It can be avoided by ultrafiltration rates greater than 300 ml/h. Backfiltration due to negative oncotic pressure has no significant consequences.

## References

1   Starling, E.H.: J. Physiol. *19:* 312−326 (1896).
2   Adair, G.S.: The thermodynamic analysis of the observed osmotic pressure of protein salts in solutions of finite concentrations Proc. R. Soc. A *126:* 16 (1929).
3   Diem, K.; Lentner, C.: Wissenschaftliche Tabellen, Documenta Geigy; 7. Ausgabe (Thieme, Stuttgart 1975).
4   Klinkmann, H.; Falkenhagen, D.; Smollich, B.P.: Durchlässigkeit von Polysulfonmembranen für Pyrogene. Investigation of the permeability of highly permeable polysulfone membranes for pyrogens (this volume, p. 174).
5   Burton, A.C.: Hemodynamics and Pysics of the circulation; in Ruch, Patton, Physiology and biophysics (Saunders, Philadelphia 1960).
6   Kuchling, H.: Taschenbuch der Physik (Harri Deutsch, Thun/Frankfurt/M.).

Dipl.-Phys. S. Stiller, Abt. Innere Medizin III, Technische Universität Aachen, Pauwelsstrasse 1, D-5100 Aachen (FRG)

Contr. Nephrol., vol. 46, pp. 33–42 (Karger, Basel 1985)

# Permeability for Middle and Higher Molecular Weight Substances

## Comparison between Polysulfone and Cuprophan Dialyzers[1]

*H. Brunner, H. Mann, S. Stiller, H.-G. Sieberth*[2]

Department of Internal Medicine II, Technical University Aachen, Aachen, FRG

## Introduction

The so-called 'middle molecule hypothesis' [1, 2] postulates the presence of substances in the sera of uremic patients in the molecular weight range between 500 and 5,000 Dalton. In the last decade, those substances have been detected in the biological fluids of patients with end stage renal failure [3]. The discussion about their possible clinical relevance stimulated besides others the development of new membranes specially designed for the elimination of this class of uremic toxins. In the present study, the usefulness of gel chromatographic methods for the characterization of a newly developed, highly permeable polysulfone hemodiafilter in comparison to a Cuprophan dialyzer is described.

## Materials and Methods

Dialysates from patients on regular hemodialysis treatment were investigated using a polysulfone hemodiafilter (Hemoflow F 60, 1.25 m², Fresenius) and a Cuprophan dialyzer (Nephross Allegro, 1.4 m², Organon Teknika). Two series of experiments were performed to study convective and diffusive transport processes, respectively. Blood flow (150 ml/min) and

[1] Supported by Deutsche Forschungsgemeinschaft (SFB 109/F 5).

[2] We thank Mrs. *Helga Riedel*, Miss *Regina Dombrowski*, Miss *Margaret Breuer*, Mrs. *Marlies Pfeiffer* and Mrs. *Annemarie Brunner* for technical assistence. We also thank the staff of our dialysis department for their cooperation.

*Table I.* Characteristics of the gel-chromatographic systems used for analysis of dialysates

| System | Column | Fractionation range[1] (for peptides)[1] | Buffer | Linear flow rate, cm/h | Detection |
|--------|--------|------------------------------|--------|------------------------|-----------|
| 1 | Sephadex G−15 1.1 × 200 cm (Pharmacia) | up to 1,500 Dalton | 0.05 mol/l NH₄HCO₃ | 75.7 | UV detector: 206 nm (Uvicord S, LKB) |
| 2 | TSK-HW (40) S 1.6 × 68 cm (Merck) | 100−10,000 Dalton | 0.05 mol/l NH₄HCO₃ | 14.9 | UV detector: 206 nm (UV−100, Varian) |
| 3 | TSK-G 2000 SW 0.75 × 60 cm with precolumn: TSK-GSWP 0.75 × 7.5 cm (LKB) | 500−60,000 Dalton | 0.1 mol/l Na₂HPO₄/ NaH₂PO₄ (pH = 6.8) 0.1 mol/l NaCl | 27.2 | 2 UV detectors in series: a) 206 nm (Uvicord SD, LKB) b) 280 nm (Varichrom, Varian) |

[1] According to instructions of manufacturers

dialysate flow (500 ml/min) were kept constant. In the first series of experiments the polysulfone and Cuprophan dialyzers were arranged in parallel in the bloodstream (for 30 min) and ultrafiltrates were collected from both dialyzers simultaneously. Afterwards, dialysis was continued using only one of the two dialysers.

In the second series, a volume of 300 ml of dialysate was recirculated for 30 min in a closed system before dialysate samples were drawn for analysis.

Additionally, during each dialysis the first 40 liters of dialysate were collected for further investigations.

Gel chromatography was performed using three different systems. The experimental details are summarized in table I and in figures 1−8.

## Results and Discussion

As the chemical nature of most 'middle' and higher molecular weight solutes in uremic sera, which have exerted great interest as possible uremic toxins, remains as yet unknown and methods for their quantitative analysis are not yet available [4−8], it seemed reasonable to establish a general overview of the whole molecular weight distribution of solutes in the dialysates from different kinds of dialysers. Gel chromatography was shown to

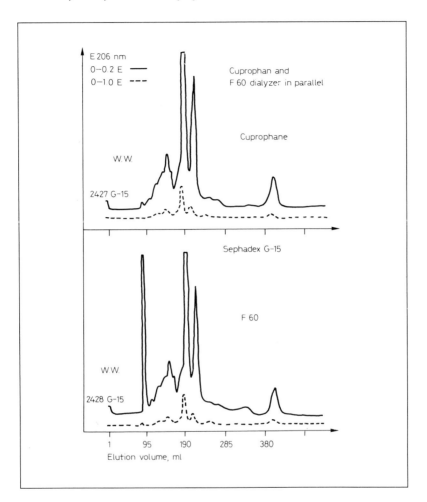

*Fig. 1.* Elution profiles of ultrafiltrates from Cuprophan and F 60 dialyzers, respectively (arranged in parallel) after gel chromatography on a Sephadex G-15 column. Injection volume 1 ml (for further details see table I, system 1).

be a sensitive method for the characterization of membranes especially when several chromatographic systems with different linear fractionation ranges were used, as summarized in table I.

In the first series of experiments, ultrafiltrates were obtained simultaneously from polysulfone and Cuprophan dialyzers, which were arranged in parallel in the bloodstream.

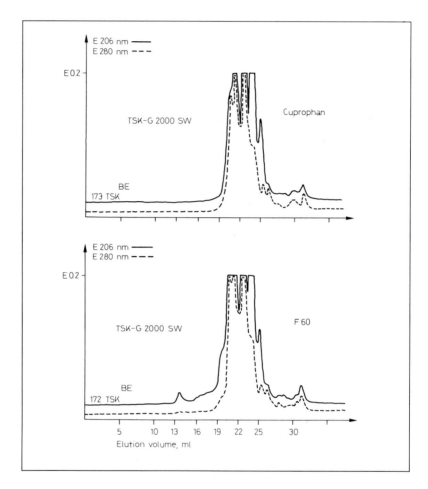

*Fig. 2.* Elution profiles of ultrafiltrates from Cuprophan and F 60 dialysers, respectively (arranged in parallel) after gel chromatography on a TSK-G 2000 SW column. Injection volume = 200 μl (for further details see table I, system 3).

As expected, the gel chromatographic profiles showed a lot of additional fractions in the 'middle' and higher molecular weight range in the Hemoflow F 60 filtrates, which could not been found in the Cuprophan filtrates (fig. 1, 2). Especially in the most sensitive system of figure 2, the corresponding regions (elution volumes between 13 and 20 ml) of the elution curves differed significantly starting with albumin as the first detectable peak.

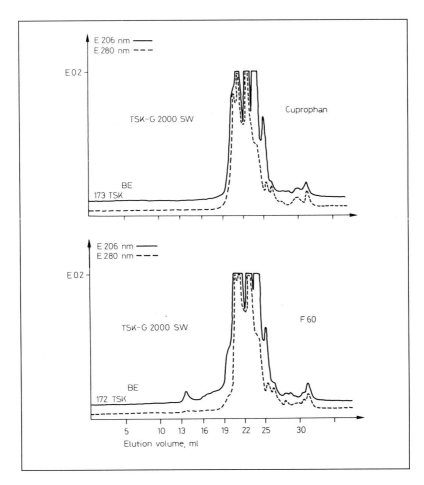

*Fig. 3.* Elution profiles of recirculated dialysates from Cuprophan and F 60 dialyzers, respectively, after gel chromatography on a TSK-G 2000 SW column. Injection volume = 200 μl (for further details see table I, system 3).

In the second experimental series, a volume of 300 ml of dialysate was recirculated for 30 min in a closed system in order to concentrate the solutes in the dialysates. As the molecular weight distributions in figure 3 show, the polysulfone dialyzer eliminated a broad spectrum of middle and higher molecular weight solutes in addition to the solutes eliminated also by the Cuprophan dialyser. Some of these differences were detectable in untreated aliquots from a 40-liter dialysate volume which was collected in the

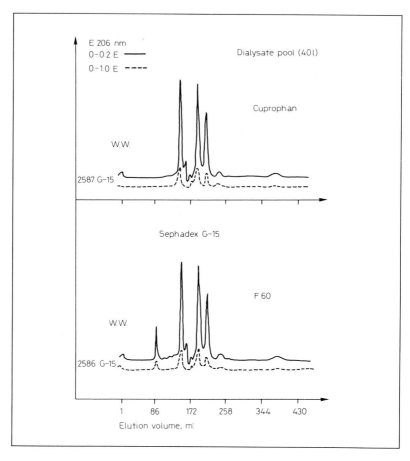

*Fig. 4.* Gel chromatography of dialysate samples drawn from a volume of 40 liters dialysate collected in the first 80 min of hemodialysis treatment. Chromatographic system: Sephadex G-15 (see table I, system 1). Injection volume = 1 ml.

first 80 min of hemodialysis treatment (fig. 4), but became more prominent after a 20-fold concentration of the dialysate by lyophilization, as shown in figure 5, where the elimination rates for middle and higher molecular weight solutes are compared.

In order to get more detailed information about the molecular sizes of the solutes which were detected only in the dialysates of the polysulfone dialyser, the column for high performance liquid chromatography (system 3 in table I) was calibrated with proteins and peptides which were run as a

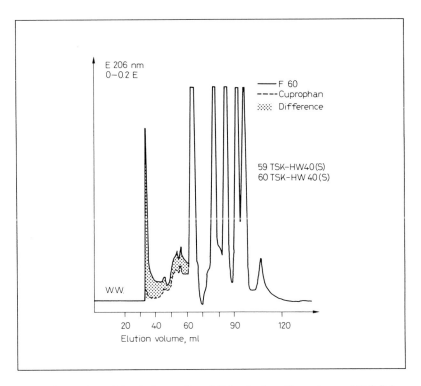

*Fig. 5.* Comparison of the elution profiles of dialysates from Cuprophan and F 60 dialyzers, respectively, after chromatography on a TSK-HW 40 (S) column. From 40 liters of dialysate which were collected in the first 80 min of hemodialysis treatment, aliquots were concentrated 20-fold by lyophilization and 100 µl injected onto the column (system 2 in table I).

mixture (fig. 6) and also individually for estimation of the linear fractionation range of the system (fig. 7). Although the interpretation of gel chromatography on calibrated columns is often misunderstood as the results depend strongly on the special experimental conditions, some conclusions can be drawn: dialysates from the Cuprophan dialyzer show an 'empty' region in their molecular weight distribution compared with dialysates from the polysulfone dialyzer. In this region, the first eluted peak was identified as albumin followed by a broad range of as yet unknown solutes with molecular weights reaching down to the region of the so-called 'middle molecules'.

The amount of albumin in the dialysates was very small and near the detection limits of the usually employed analytical methods. Therefore, a

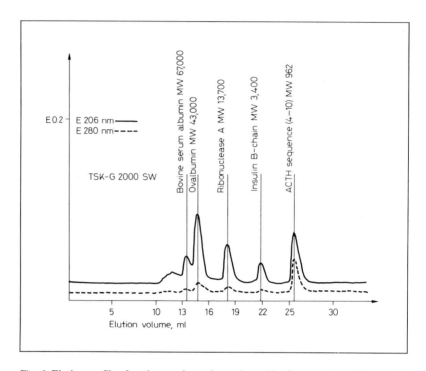

*Fig. 6.* Elution profile of a mixture of proteins and peptides (between 4 and 13 µg each) after chromatography on a TSK-G 2000 SW column. Injection volume = 20 µl (for further details see table I, system 3).

rough quantitative estimation was derived from the elution curves of gel chromatography after calibration of the column with known amounts of serum albumin (fig. 8). A great advantage of the method is that interfering solutes are separated from the interesting peak whose extinction is measured at 206 nm. So it could be shown that the collected dialysate volume of 40 liters contained less than 0.4 g albumin.

Summarizing the results, it was found that gel chromatography is a useful tool for the characterization of the molecular weight distributions in dialysates from different dialyzers. The described polysulfone dialyser eliminates a broad spectrum of 'middle' and higher molecular weight solutes which are not eliminated by the Cuprophan dialyser while the losses of albumin are very small. Further investigations should be performed in the future to answer the question of the possible clinical relevance of these findings.

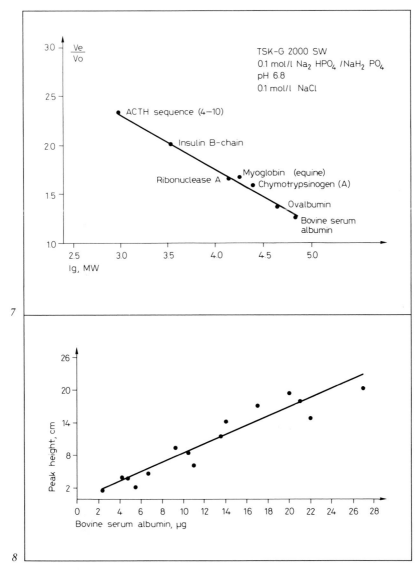

*Fig. 7.* Determination of the linear fractionation range for proteins and peptides of the TSK-G 2000 SW column. $V_e$ = Elution volume of a substance; $V_o$ = exclusion volume of the system. All substances were run individually. Experimental details as in figure 6.

*Fig. 8.* Calibration curve for the quantitative estimation of microgram amounts of albumin in biological fluids using fractogel TSK-HW 40 (S) and detection at 206 nm. Injection volume = 100 μl with different amounts of serum albumin as indicated on the x-axis (for further details see table I, system 2).

## References

1    Babb, A.L.; Popovich, R.P.; Christopher, T.G.; Scribner, B.H.: The genesis of the
     square meter-hour hypothesis. Trans. Am. Soc. artif. internal Organs *17:* 81–91 (1971).
2    Babb, A.L.; Farrell, P.C.; Uvelli, D.A.; Scribner, B.H.: Hemodialyzer evaluation by
     examination of solute molecular weight spectra: implications for the square meter-hour
     hypothesis. Trans. Am. Soc. artif. internal Organs *18:* 98–105 (1972).
3    Brunner, H.; Mann, H.: «Mittelmoleküle» als Urämietoxine; in Losse, Renner,
     Klinische Nephrologie, pp. 489–501 (Thieme, Stuttgart 1982).
4    Bergström, J.; Fürst, P.: Uraemic middle molecules. Clin. Nephrol. *5:* 143–152 (1976).
5    Bergström, J.; Funck-Brentano, J.L.; Klinkmann, H.: The road towards the identifica-
     tion of middle molecules. Artif. Organs *4:* 209–210 (1980).
6    Bergström, J.; Fürst, P.: Uraemic toxins; in Drukker, Parsons, Maher, Replacement of
     renal function by dialysis, pp. 354–390 (Nijhoff, Boston 1983).
7    Contreras, P.; Later, R.; Navarro, J.; Touraine, J.L.; Freyria, A.M.; Traeger, J.:
     Molecules in the middle molecular weight range. Nephron *32:* 193–201 (1982).
8    Schoots, A.C.; Mikkers, F.E.P.; Claessens, H.A.; De Smet, R.; Van Landschoot, N.;
     Ringoir, S.M.G.: Characterization of uremic 'middle molecular' fractions by gas
     chromatography, mass spectrometry, isotachophoresis, and liquid chromatography.
     Clin. Chem. *28:* 45–49 (1982).

Dr. rer. nat. H. Brunner, Technische Hochschule Aachen,
Department Innere Medizin II, Goethestrasse 27–29, D-5100 Aachen (FRG)

Contr. Nephrol., vol. 46, pp. 43–60 (Karger, Basel 1985)

# Optimising of Hemodiafiltration with Modern Membranes?

*K.G.B. Sprenger[a], H. Stephan[a], W. Kratz[b], K. Huber[b], H.E. Franz[a]*

[a]Section of Nephrology, Department of Medicine I; [b]Department of Mathematics V, University of Ulm, FRG

## Introduction

Owing to its excellent solute elimination and treatment tolerance, hemodiafiltration (HDF) has become an established alternative blood purification procedure. By 1979 in Japan, 134 patients had already undergone hemodiafiltration in 108 centers [5].

Based on more recent findings [2, 7–9] uremic toxins are also suspected in the large molecular range of 10,000–50,000 daltons. So far, the high-flux dialyzers had only a very low clearance effectivity for molecules in this range. Up to now, the sieving coefficient for vitamin A retinol-binding proteins with a molecular weight of 21,000 daltons was well below 0.1 for all dialyzers [8].

Consequently, the investigation of the following questions became obvious: (1) What are the effects of diffusive and convective clearance components upon the total elimination effectivity in HDF? (2) Which dialyzer properties are required for HDF? (3) Can HDF be optimised with protein-permeable membranes or can we perhaps do without HDF in the future?

## Methods

We determined the clearances for the following 7 substances (urea, creatinine, uric acid, phosphate, glucose, vitamin $B_{12}$ and inulin) for 4 dialyzers (2 capillary dialyzers: PAN 15 and Filtryzer B 1 M; 2 plate dialyzers: RP 610 and Lundia major High-Flux) during hemodialysis (HD) and HDF.

Six tests were performed for each dialyzer and each procedure and the clearances were determined at four times (after 5, 50, 210 and 240 min). Before blood sampling in HD, the ultrafiltration was interrupted for 10 min; in HDF the filtrate flow was measured for a period of 10 min. Before and after the HDF sessions, an additional hemofiltration was performed and the clearance and sieving coefficients for all substances were determined after 10 min. Uniform test conditions prevailed for all procedures: blood flow 200 ml/min, transmembrane pressure 500 mm Hg, dialysate flow in HD and HDF 500 ml/min. For calculation of the plasma flow, hematocrit was measured in the first and last blood samples.

## Clearance Measurements

Physiological concentrations of vitamin $B_{12}$ are 100% bound to plasma protein. To perform a clearance determination, it must therefore be supplied in a sufficient quantity to ensure that the protein transport capacity is exceeded and the an adequate proportion of free vitamin $B_{12}$ is available during the entire time of treatment.

In order to reach the necessary plasma levels before the onset of the treatment, 1 mg of cyanocobalamin and 5 mg of sinistrin (Inutest®) were administered by bolus injection. The same volume was administered via a perfusion pump into the blood as a 'maintenance dose' during the treatment.

Vitamin $B_{12}$ analysis was made with a RIA (Phadebas), sinistrin and glucose with an en-enzymatic analysis using a test set by Boehringer Mannheim, inorganic phosphate was determined with the molybdenum blue method, and urea, creatinine and uric acid were determined with Beckman analyzers.

## Clearance Calculations

The derivations of the following formulas are explained in detail elsewhere [4, 11, 12]. HD clearance:

$$K_{HD} = Q_B \, \frac{C_{Bi} - C_{Bo}}{C_{Bi}} \, ,$$

$Q_B$ = Blood flow; $C_{Bi}$ = blood concentration at dialyzer inlet; $C_{Bo}$ = blood concentration at dialyzer outlet. HF clearance:

$$K_{HF} = Q_F \, \frac{C_F}{C_{Bi}} \, ,$$

$Q_F$ = filtrate flow; $C_F$ = filtrate concentration. HDF clearance:

$$K_{HDF} = Q_B \frac{C_{Bi} - C_{Bo}}{C_{Bi}} + Q_F \, \frac{C_{Bo}}{C_{Bi}} \quad \text{or} \quad K_{HDF} = R_x \, (Q_F \times S + P),$$

$R_x$ = average concentration relation $\dfrac{C_B}{C_{Bi}}$ ; S = sieving coefficient; P = permeability.

Diffusive proportion of HDF clearance: $K_{HDF_D} = R_x \times P$ with the permeability resulting from:

$$P = Q_F \left( 1 + \frac{\ln (C_{Bo}/C_{Bi})}{\ln (1 - Q_F)/Q_{Bi}} - S \right) ,$$

and the average concentration relation $\dfrac{C_B}{C_{Bi}} = R_x$ from:

$$R_x = \frac{Q_{Bi}}{Q_F \times B}\left(1-\left(1-\frac{Q_F}{Q_{Bi}}\right)^B\right)$$

and the auxiliary quantity B from

$$B = 1 + \frac{\ln(C_{Bo}/C_{Bi})}{\ln(1-Q_F)/Q_{Bi})} \ .$$

Convective proportion of HDF clearance: $K_{HDF_K} = Rx \times Q_F \times S$;
plasma clearance: $Q_P = Q_B(1-HK)$. $Q_p$ = Plasma flow; HK = Hematocrit.

For the calculation of the phosphate, vitamin $B_{12}$ and inulin clearances, the plasma flow was entered into the respective clearance equation instead of the blood flow, since due to the low mass transfer coefficient before the blood cell separation, no concentration compensation occurs between erythrocytes and plasma which is worth mentioning [11, 12].

### 2-Pool Model

The recently described [11] 2-pool model was used for the treatment simulations. The variables and values used are compiled in table I. The dialysis index (DI) was calculated as follows [14]:

$$DI = \frac{\dfrac{n \times h}{Wo}\,K + 10080\,K_R}{(30\,L/Wo)\,(KO/1.73)},$$

n = number of dialyses; h = hours of dialysis; Wo = week; K = dialyzer clearance (ml/min); 10080 $K_R$ = residual renal clearance (ml/min)/week; KO/1.73 = standard body surface per m².

### Results and Discussion

#### 1. What are the effects of diffusive and convective clearance components upon the total elimination effectivity in HDF?

In HDF the mass transport of solutes across the membrane consists of a diffusive and a convective component, both of which have a marked influence upon the dialyzer clearance.

The *diffusive* component is of primary importance for the elimination of small molecules, which would require a very high filtrate flow if attempted in a convective process alone. The product of membrane permeability and surface as well as the blood and dialysate compartment flow geometry of the dialyzer are decisive for the solute transport. The convective component plays an important role in the transport of larger molecules, where the membrane permeability diminishes rapidly with increasing molecular size.

*Table I.* Variables in the 2–pool simulation

| Variables | HDF | HD | HDF | HDF |
|---|---|---|---|---|
| Filtration rate, F (ml/min) | 65 | 0 | 50 | 100 |
| Dialyser clearance, K (ml/min) | | | | |
| RP 6[1]: Creatinine | 128 | | | |
| Inulin | 76 | | | |
| F 60[2]: Creatinine | | 168 | 173 | 178[3] |
| Inulin | | 85 | 93 | 101[3] |
| $\beta_2$-microglobulin | | 56 | 81 | 106[3] |
| Mass transfer coefficient T (ml/min) | | | | |
| Creatinine[4] | 150 | 150 | 150 | 150 |
| Inulin[5] | 3,4 | 3,4 | 3,4 | 3,4 |
| $\beta_2$-microglobulin[6] | 2,5 | 2,5 | 2,5 | 2,5 |
| Residual renal function, $K_R$ (ml/min) | 0 | 0 | 0 | 0 |
| Generation rate, G (ml/min) | 0 | 0 | 0 | 0 |
| Body weight, KG (kg) | 70 | 70 | 70 | 70 |
| Total body volume, V (l); $V^4 = 0{,}57 \times KG$ | 52,5 | 52,5 | 52,5 | 52,5 |
| Intracellular space, IZR (l); $IZR^4 = 2/3\ V$ | 35 | 35 | 35 | 35 |
| Extracellular space, EZR (l); $EZR^4 = 1/3\ V$ | 17,5 | 17,5 | 17,5 | 17,5 |
| Distribution coefficient, $X^4$ | 1 | 1 | 1 | 1 |

[1] From *Kohnle* et al. [3].
[2] From *Streicher and Schneider* [13].
[3] Linear extrapolation based on studies by *Dieter* [1].
[4] From *Sprenger* et al. [11]
[5] From *Popovich* et al. [6].
[6] Estimated according to *Popovich* et al. [6].

The effectivity of the convective solute transport is determined by 2 parameters, namely filtration rate and sieving coefficient.

*The total elimination effectivity* of HDF does not equal the total of the two single procedures, since diffusion and convection do not add up but take place simultaneously and mutually affect each other. Figure 1 shows

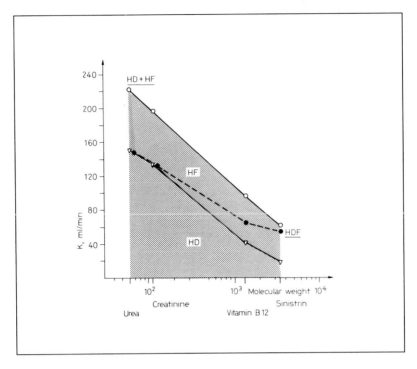

*Fig. 1.* Comparison of hemodiafiltration clearances with the summation of hemodialysis and hemofiltration clearance values.

the HD, HF and HDF clearances for the RP 610, determined under similar test conditions. The upper line represents the summation of the HD and HF clearance values, the dashed line links the clearance values measured in HDF. From a measuring aspect, no explanation of this interaction is possible since the filtrate concentration of a substance cannot be determined in HDF due to the mixing of concentrate and dialysate. The equation which we used to calculate the clearance does not provide insight into the interaction between convection and diffusion [10].

The mathematical model developed by *Lewis* [4] permits division of the total clearance into the convective and the diffusive components. After calculation with the help of this model, the relative proportion of convection in the total urea clearance amounts only to 28%, but to 67% for inulin. The absolute proportion of the convective clearance component remains about constant over the entire molecular weight range, whereas the diffu-

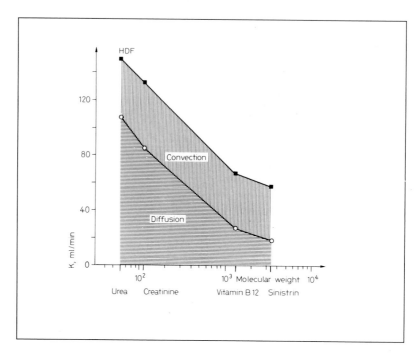

*Fig. 2.* Diffusive and convective clearance components in hemodiafiltration (shown for the RP 610 dialyzer).

sive clearance proportion drops rapidly (fig. 2). This can be attributed to the fact that the decrease of the permeability (fig. 3a) associated with increasing molecular weight occurs to a much greater extent than the sieving coefficient (fig. 3b). Furthermore, the upper graph also shows the marked inhibition of the diffusion — especially for smaller molecules — caused by the superimposed convection in HDF. The permeability (here RP 610) for substances up to a molecular weight of vitamin $B_{12}$ is more than 40% lower in HDF; however, the permeability for inulin is only 14% lower than in HD.

Moreover, our studies showed that the clearance during HDF treatment decreased much more markedly than in HD, as shown in the example of a vitamin $B_{12}$ clearance achieved with the PAN 15 (fig. 4). Here the decrease in the convective solute transport is caused by a drop in the filtration rate (fig. 5a) and sieving coefficient (fig. 5b) due to protein deposits upon the membrane. The diffusive solute transport is also impaired by the secondary membrane; however, to a much lesser extent. With the PAN 15,

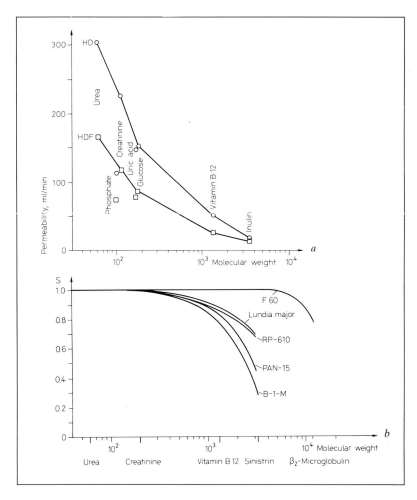

*Fig. 3. a* Permeability versus molecular weight in hemodialysis (HD) and hemodiafiltration (HDF), *b* Sieving coefficient versus molecular weight for 5 high-flux dialyzers (measured at the onset of the treatment; for F 60, values from *Streicher and Schneider* [13] are used).

the vitamin $B_{12}$ permeability decreases only by 15%; however, the filtration rate decreases by 30% and the sieving coefficient by 40%.

When analyzing the effects of the solute transport parameters upon the two clearance components, it is evident that the decrease in the total clearance is the exclusive responsibility of the convective component. The

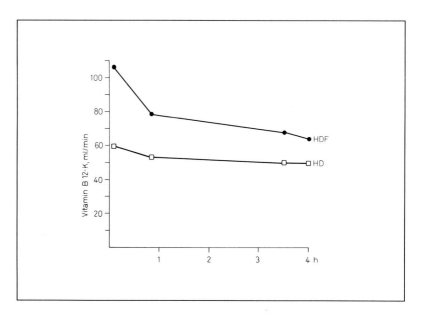

*Fig. 4.* Course of vitamin B$_{12}$ clearance during hemodialysis (HD) and hemodiafiltration (HDF).

diffusive component even increases since the convection-induced impairment of the diffusion is markedly reduced during the course of the treatment (fig. 6).

Thus, the calculation models revealed that convection leads to a marked impairment of the diffusion in HDF. This explains on one hand the influence of HDF upon the clearance of small molecules – which is low in comparison to HD –, and on the other hand an increase of the diffusive clearance component with a decreasing total clearance.

### 2. Which dialyzer properties are required for HDF?

Only high-flux dialyzers can be used for HDF, and those must meet special requirements such as high hydraulic permeability and high cut off in order to eliminate large quantities of middle and large molecules. Furthermore, they must have good diffusive properties under HDF conditions.

Our studies comparing capillary and plate dialyzers in HD and HDF showed that aside from permeability and surface, the type of the membrane

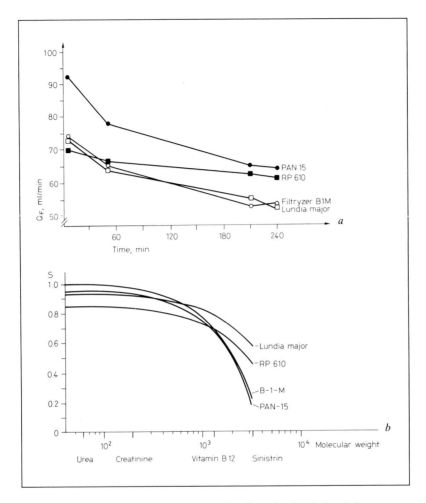

*Fig. 5. a* Course of filtration rate in hemodiafiltration using 4 high-flux dialyzers.
*b* Sieving coefficient versus molecular weight for 4 high-flux dialyzers (measured at the end of treatment).

also plays a decisive role in the diffusive performance of a hemodiafilter. With hollow fibre dialyzers, higher urea and creatinine clearances are achieved in HDF (fig. 7a) whereas for plate dialyzers this is applicable to HD (fig. 7b). Since the diffusion impairment caused by the convection is similar for all dialyzers, the diffusion must be further negatively influenced

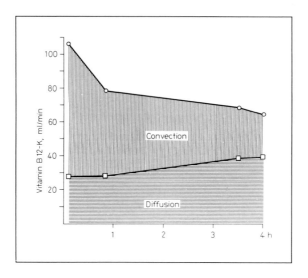

*Fig. 6.* Time course of total clearance and diffusive clearance proportion in hemodiafiltration (example: vitamin $B_{12}$ clearance, PAN dialyzer).

by HDF for plate dialyzers. The reason for this is the relatively high transmembrane pressure which leads to the following changes in the plate dialyzers: (1) Expansion of the blood compartment: expanding the diffusion distance which has to be passed by the solutes. (2) Membrane is partly pressed upon the supporting structure: leading to a reduction in active membrane surface area. (3) Partial unfolding of the membrane layers on the dialysate side minimises the dialysate compartment and creates a channelling of the dialysate flow.

This *plate effect* also applies to higher molecular substances. Here, however, the performance-increasing influence of the convective transport dominates over the reduction in the diffusive clearance component so that the total clearance in HDF increases.

*Fig. 7.* Clearance versus molecular weight in hemodiafiltration (HDF) and hemodialysis (HD). *a* With 2 capillary dialyzers: HDF clearances for all substances higher than HD clearances. *b* With 2 plate dialyzers: HDF clearances for urea and creatinine lower than HD clearances.

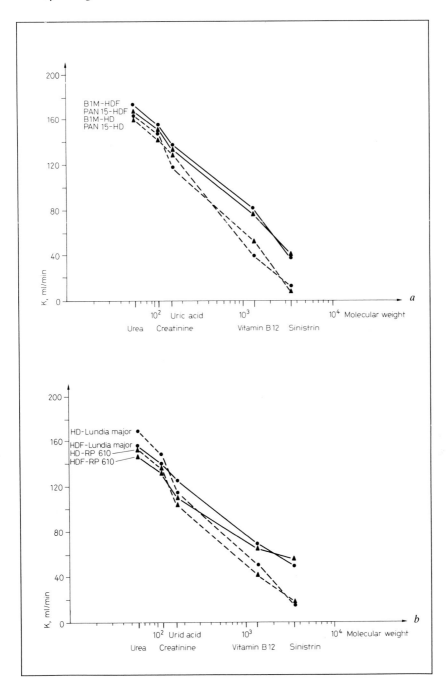

In addition to favorable diffusion and convection properties, a dialyzer suitable for HDF must also have a stable blood and dialysate flow structure.

*3. Can HDF be optimised with protein-permeable membranes or can we perhaps do without HDF in the future?*

To clarify this question, we made the following comparisons based on 2-pool model simulations: (1) A conventional HDF (for example, with the RP 6 dialyzer) with a modern HDF (for example with the F 60 polysulfone hemodiafilter). (2) HDF with F 60 with the presently used filtrate flow of 50 ml/min with a filtrate flow of 100 ml/min which will be possible in the future. (3) HD with F 60 and HDF with F 60.

The toxin markers we used were creatinine for small molecules, inulin for middle molecules and $\beta_2$-microglobulin for large molecules. The dialysis index of each of these substances was calculated for the different treatment modes as a guideline for the minimum clearance performance sufficient for 1 week. The figures show the treatment time and the decrease in concentration at which for an average patient without residual kidney function a dialysis index of 1 is reached with three weekly treatment sessions.

Membranes which can be used also for the removal of large molecules from the blood have only recently become available. Whereas previously the clearance efficiency in the large molecular range was practically negligible, even in HDF (fig. 8a), now protein-permeable membranes achieve under standard conditions ($Q_B$ 200 ml/min, $Q_D$ 500 ml/min, $Q_F$ 50 ml/min) a clearance of large molecules which approaches the renal glomerular membrane (fig. 8b).

The advantage of the improved small and middle molecular clearance is evident by the difference in the concentration decrease in the intra- (IZR) and extracellular space (EZR) which takes place during dialysis with one of these dialyzers. With the F 60, an extracellular creatinine level is reached after 3 h, and after 4 h the same inulin concentration as with the RP 6 after 5 h (fig. 9a, b) is achieved. In the intracellular space, the creatinine level is reduced by an additional 7% within 6 h (fig. 9a). The intracellular middle and large molecular concentration changes during HDF only by 1−2% due to its slow transfer (fig. 9b). Here the higher clearance leads to a significant difference only after the end of the treatment or after several treatment sessions. Due to its high hydraulic permeability, the effectivity of the F 60 is fully utilised only when the filtration rate is increased from 50 to 100 ml/min. As expected, the small molecular clearance remained almost constant and the middle molecular clearance increased

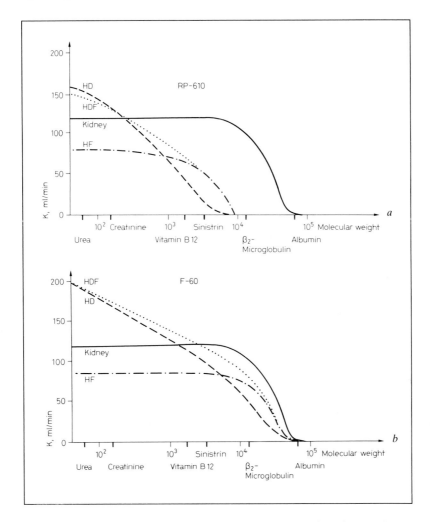

*Fig. 8.* Clearance versus molecular weight in hemodiafiltration (HDF), hemodialysis (HD), hemofiltration (HF) and kidney. *a* With RP 610 dialyzer. *b* With F 60 hemodiafilter.

only by approximately 10% (fig. 10a, b). However, the large molecular clearance is increased by 24% so that with the same elimination quantities the treatment time can, for example, be reduced from 4 to 3 h (fig. 10c).

When comparing this F 60-HDF with F 60-HD, the increase in the creatinine clearance is also insignificant (fig. 11a). However, the much higher inulin clearance leads to a 7% difference in the extracellular level

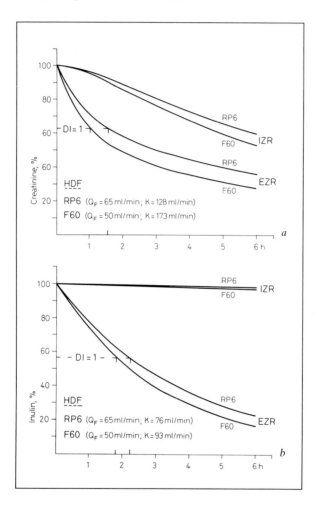

*Fig. 9.* Comparison of concentration decrease in intra- and extracellular space in hemodiafiltration with RP 6 and F 60 dialyzers. *a* Creatinine. *b* Inulin.

*Fig. 10.* Comparison of concentration decrease in hemodiafiltration (HDF) with a F 60 hemodiafilter at a blood flow of 50 ml/min (HDF$_1$) and 100 ml/min (HDF$_2$). *a* Creatinine. *b* Inulin. *c* β$_2$-microglobulin.

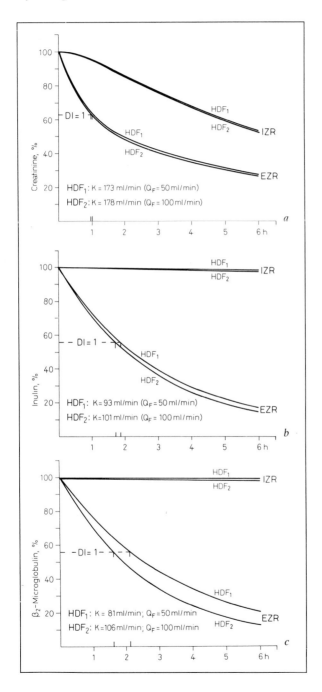

after 3 h (fig. 11b). Apparently, the decisive advantage of HDF is in the $\beta_2$-microglobulin clearance which amounts to twice the HD clearance. Here the extracellular concentration is 23% lower after 3 h than in HD (fig. 11c).

Thus, our model simulations showed that we cannot do without HDF in the future. Perhaps the protein-permeable membranes will permit an optimising of HDF in the large molecular range, since with a filtrate flow of 100 ml/min the large molecular clearance can be adapted to the middle molecular clearance (table I). The dialysis index for inulin and $\beta_2$-microglobulin will then be above 1 with a weekly treatment duration of $3 \times 2$ h.

Therefore, we consider a weekly HDF period of 9 h to be sufficient for every patient. In this case, the substitution volume amounts to 18 liters in 3 h, corresponding to the extracellular volume in the average patient. During 3 h of treatment, the extracellular concentrations of creatinine, inulin and $\beta_2$-microglobulin drop by 60, 64 and 66%, respectively, and the dialysis index for inulin increases to 1.8 and to 1.9 for $\beta_2$-microglobulin. An extension of the treatment time achieves only insignificant changes in the toxin levels due to the decreasing effectivity.

## Outlook

It has been proven that, for the present time, large molecules can be most effectively eliminated with HDF [9, 13]. A first report outlining 1 year's experience with a protein-permeable membrane [9] describes a significant hematocrit increase in HDF patients. This seems to confirm the assumption that the accumulation of large molecular substances in the body may have great importance in the pathogenesis of uremia. Therefore, it should be hoped that with HDF progress in the treatment of uremia will be possible in the future. However, we should not forget that the $\beta_2$-microglobulin clearance of the normal kidney is in the range of 1,000 liters per week and thus still amounts to 20-fold of the HDF clearance which can be realised with the modern membranes.

---

Fig. 11. Comparison of concentration decrease with a F 60 dialyzer in hemodialysis (HD) and hemodiafiltration (HDF). a Creatinine. b Inulin. c $\beta_2$-microglobulin.

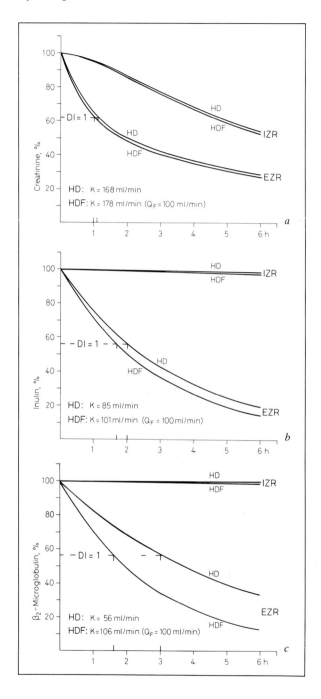

## References

1   Dieter, K.: Stofftransport bei Hämodialyse, Hämofiltration und Hämodiafiltration; in Schütterle, Wizemann, Seyffart, Hemodiafiltration, p. 7 (Hygieneplan, Oberursel 1982).

2   Jørstad, S.; Smeby, L.C.; Widerøe, T.-E.; Berg, K.J.: Transport of uremic toxins through conventional hemodialysis membranes. Clin. Nephrol. *12:* 168 (1979).

3   Kohnle, W.; Sprenger, K.B.G.; Spohn, B.; Franz, H.E.: Hemodiafiltration using readily available equipment. J. Dial. *1:* 27 (1979).

4   Lewis, A.E.D.: Mathematical modelling of hemodiafiltration with particular reference to prediction of clearance; in Schütterle, Wizemann, Seyffart, Hemodiafiltration, p. 63 (Hygieneplan, Oberursel 1982).

5   Maekawa, M.; Kishimoto, T.; Ohyama, T.; Tanaka, H.: Present status of hemofiltration and hemodiafiltration in Japan. Artif. Organs *4:* 85 (1980).

6   Popovich, R.P.; Hlavinka, D.J.; Bomar, J.B.; Moncrief, J.W.; Decherd, J.F.: The consequences of physiological resistances on metabolite removal from the patient-artificial kidney system. Trans. Am. Soc. artif. internal Organs *21:* 108 (1975).

7   Rabin, E.Z.; Algom, D.; Freedman, M.H.; Geunther, L.; Dardick, I.; Tattrie, B.: Ribonuclease activity in renal failure. Nephron *27:* 254 (1981).

8   Röckel, A.; Gilge, U.; Liewald, A.; Heidland, A.: Elimination of low molecular weight proteins during hemofiltration. Artif. Organs *6:* 307 (1982).

9   Saito, A.; Ohta, K.; Takay, T.; Chung, T.G.: One year experience of protein-permeating hemodiafiltration. Artif. Organs *7:* 58 (1983).

10  Sprenger, K.B.G.: Hemodiafiltration. Life Supp. Systems *1:* 127 (1983).

11  Sprenger, K.B.G.; Kratz, W.; Lewis, A.E.; Stadtmüller, U.: Kinetic modeling of hemodialysis, hemofiltration and hemodiafiltration. Kidney int. *24:* 143 (1983).

12  Stephan, H.-G.: Hämodiafiltration und Hämodialyse: Klinischer Test mit vier Dialysatoren; Diss. Ulm (1982).

13  Streicher, E.; Schneider, H.: Stofftransport bei Hämodiafiltration. Nieren- Hochdruck-Krankh. *12:* 339 (1983).

14  Techan, B.P.; Gacek, E.M.; Heymach, G.J.; Brown, J.; Smith, L.J.; Sigler, M.H.; Gilgore, G.S.; Schleifer, C.R.: A clinical appraisal of the dialysis index. Trans. Am. Soc. artif. internal Organs *23:* 548 (1977).

Dr. K.B.G. Sprenger, Dept. of Nephrology, University of Düsseldorf, Moorenstraße 5, 4000 Düsseldorf 1 (FRG)

Contr. Nephrol., vol. 46, pp. 61–68 (Karger, Basel 1985)

# Removal of Hormones by Hemofiltration and Hemodialysis with a Highly Permeable Polysulfone Membrane

*V. Wizemann[a], H.G. Velcovsky[a], H. Bleyl[b], S. Brüning[a], G. Schütterle[a]*

Departments of [a]Internal Medicine and [b]Clinical Chemistry, University of Giessen, FRG

## Introduction

For the last two decades, chronic dialysis has been a successful therapy for maintaining the life of patients suffering from end stage renal failure.

Yet despite undeniable progress in safety and comfort of dialysis treatment, the present dialysis regime primarily keeps patients alive and does not assure a satisfactory compensation of physiological kidney deficiency. Although progress in renal replacement includes the development of new membranes with an improved biocompatibility and higher sieving coefficients for substances with molecular weights of 100–5,000 daltons, one cause for the failure of dialysis therapy to simulate complete renal function may lie in its inability to remove, in addition, solutes of higher molecular weight. The kidney is an important organ in the catabolism of small proteins, polypeptides, and polypeptide hormones. So far, available membranes for dialysis and hemofiltration therapy are unable to remove larger quantities of those peptides. It is the aim of the present study to evaluate the properties of a highly permeable polysulfone membrane and to study its clinical effects over a period of 4 months, where a routine dose of dialysis is prescribed. Since the nature of the molecular size of uremic toxins is not yet known in detail, peptide hormones with different molecular weights are used as substitutes for still unidentified toxins.

## Patients and Methods

In 10 patients (age 27–69 years, 4 females, 6 males) sieving coefficients of the polysulfone hemodiafilter (Hemoflow F 60, 1.25 m² capillary filter, Fresenius AG, Bad Homburg, FRG) were tested during a single hemofiltration. Sieving coefficients ($S_C$) were calculated as $S_C = 2 \times$ concentration in the filtrate/concentration in the blood inlet + concentration in the blood outlet.

Creatinine, inulin, amylase and albumin were determined by routine techniques using commercial kits. Parathyroid hormone (PTH) was measured by a C-terminal assay (Diagnostic Systems Lab. Inc., Webster, USA) and $C_{44-68}$ fragments by the assay of Henning GmbH, FRG. $\beta_2$-microglobulin (beta-2-MG, Phadebas-beta-2 micro test, Pharmacia GmbH, FRG), insulin (Phadeseph Insulin RIA, Pharmacia GmbH, FRG), thyrotropin (TSH, Behringwerke AG, FRG), and prolactin (Serono GmbH, FRG) were determined by radioimmunoassay and retinol-binding protein by immunodiffusion (LC-Partigen, Behringwerke AG, FRG).

In 10 male patients (age 34–55 years) who had been treated by chronic dialysis 3 × 4 h/week for at least 0.5 years using Cuprophan membranes (Hemoflow D 3 and D 6, Fresenius AG, Bad Homburg, FRG), the effect of the polysulfone membrane was evaluated during a 4-month period. Treatment time was kept constant (12 h/week) as well as blood flow (200–300 ml/min), dialysate flow (500 ml/min) and treatment by the same type of dialysis machine (A2008C, Fresenius AG). Ultrafiltration was controlled volumetrically and the possibility of backfiltration of dialysate through the permeable membrane into the blood was ruled out by a positive pressure in the blood compartment, leading to an ultrafiltration exceeding 600 ml/h. Urea, creatinine, phosphate, cholesterol, total protein, albumin, pseudocholinesterase and hemoglobin were measured each month after a 3-day interval of dialysis.

## Results

Sieving coefficients (Sc) of the F 60 polysulfone membrane are given in figure 1. Substances with a molecular weight of less than 5,000 daltons can permeate the membrane without being retained. Assuming that peptide substances like PTH, $\beta_2$-microglobulin, prolactin and amylase are not proteinbound, they can be used as molecular markers. Thus, there is still a small but measurable Sc for amylase with a molecular weight of 57,000 daltons. However, even after concentrating the filtrate by factor 5, albumin could not be detected in the filtrate.

During hemofiltration with the F 60 there is a small but consistent decrease in the plasma concentration of prolactin (fig. 2). However, 44 h after the treatment, the pretreatment level is again reached. Although TSH is only a slightly larger molecule when compared to prolactin, plasma concentrations did not change during hemofiltration. Accordingly, we could not

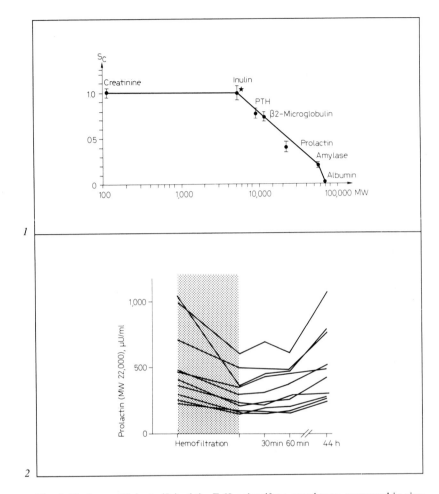

*Fig. 1.* Sieving coefficients ($S_C$) of the F 60 polysulfone membrane, measured in vivo during hemofiltration.

*Fig. 2.* Prolactin plasma levels during hemofiltration with a polysulfone membrane (F 60) and during 44 h after treatment.

measure TSH in the filtrate, indicating that the hormone of a hormone-protein complex cannot penetrate the polysulfone membrane.

Insulin in the plasma varied over a large scale of concentration during and after hemofiltration.

In some patients there was a distinct increase in insulin concentration, due to a glucose load from the substitution fluid or breakfast. In 1 diabetic

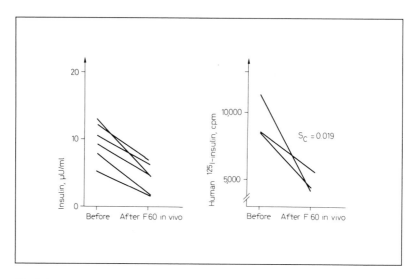

*Fig. 3.* Insulin blood levels before and after contact with the polysulfone membrane in vivo (left) and in vitro (right).

patient starting with super-normal insulin concentrations form exogenous application, a sharp decrease in circulating insulin levels during hemofiltration occurred. Insulin measured in the blood before and after contact with the polysulfone membrane showed a concentration gradient. Since insulin could not be detected in the filtrate immunologically, we perfused labelled human insulin into the blood compartment of the hemodiafilter in vitro.

The finding of decreased [125]I-insulin concentrations after the filter (fig. 3) and the fact that insulin was nearly unable to permeate the membrane supports the theory that the hormone might be adsorbed onto polysulfone. C-terminal fragments of PTH were supernormally elevated in all patients. Due to a sieving coefficient of less than 1, concentrations of PTH fragment in the hemofiltrate were lower when compared to plasma (fig. 4).

The second group of patients, who were investigated over a period of 4 months hemodialysis with a polysulfone dialyzer showed neither an improvement nor a deterioration of the biochemical parameters measured when compared to the preceding period of dialysis with Cuprophan dialysers (table I). There was neither an improvement of renal anemia nor an indication of a protein loss.

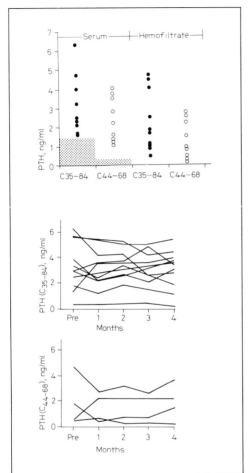

*Fig. 4.* C-terminal PTH fragments in the serum and hemofiltrate during hemofiltration with the polysulfone membrane (top). Course of PTH in the serum during 4 months of hemodialysis with the F 60 hemodiafilter (bottom).

However, the surface area of the F 60 hemodiafilter is 20–40% smaller than that of the compared Cuprophan dialyzers Hemoflow D 3 and D 6.

Plasma concentrations of both measured C-terminal PTH fragments did not change during the treatment period with the polysulfone membrane (fig. 4). The same result was obtained in 2 patients who were treated by chronic hemofiltration with the F 60 over a period of 3 months.

The membrane was tolerated well by all patients. There were no fever reactions and no signs of first-use syndrome.

*Table I.* Biochemical results of 10 patients (second group) dialyzed with Cuprophan dialysers (Hemoflow D 3 and D 6) and thereafter with a polysulfone dialyzer (Hemoflow F 60)

| | Control | 1 | Months on Hemoflow F 60 2 | 3 | 4 |
|---|---|---|---|---|---|
| Urea, MG/DL | 156 ± 36 | 167 ± 42 | 163 ± 37 | 179 ± 45 | 153 ± 23 |
| Creatinine, MG/DL | 13.6 ± 1.9 | 14.8 ± 2.0 | 13.3 ± 1.6 | 13.8 ± 1.8 | 12.0 ± 1.9 |
| Phosphate, MG/DL | 5.6 ± 1.2 | 5.8 ± 1.9 | 5.5 ± 1.6 | 5.5 ± 1.3 | 5.7 ± 1.8 |
| Cholesterol, MG/DL | 215 ± 50 | 208 ± 44 | 191 ± 48 | 194 ± 40 | 216 ± 55 |
| Total protein, G/L | 72 ± 8 | 74 ± 10 | 68 ± 6 | 69 ± 5 | 74 ± 6 |
| Albumin, G/L | 42 ± 6 | 43 ± 5 | 38 ± 5 | 41 ± 5 | 45 ± 6 |
| Pseudo-CHE, U/L | 3,342 ± 679 | 3,331 ± 792 | 3,370 ± 481 | 3,337 ± 599 | 3,668 ± 689 |
| Hemoglobin, G/DL | 9.8 ± 2.1 | 9.4 ± 2.1 | 9.5 ± 2.4 | 9.1 ± 1.9 | 9.9 ± 2.5 |
| Amylase, U/L | 78 ± 31 | 78 ± 27 | 73 ± 24 | 78 ± 32 | 80 ± 35 |
| Retinol-Binding protein, MG/DL | 15.9 ± 5.8 | 18.3 ± 2.5 | 16.1 ± 5.9 | 15.9 ± 4.1 | 16.6 ± 4.8 |

## Discussion

The membrane characteristics of the polysulfone membrane F 60 differ clearly from Cuprophan, polymethylmethacrylate, polycarbonate and polyacrylonitrile used for routine hemodialysis and hemofiltration. However, one should be cautious in interpreting sieving coefficients of peptide substances as long as protein binding and its possible determinants cannot be defined in uremia as well as during the passage through a dialyzer. Using dextrans of different molecular size as markers, *Henderson* et al. [1] demonstrated in vitro that the F 60 membrane excels other types of membranes in simulating the sieving coefficients of the animal and human kidney, which is in accordance with our findings. In the higher molecular range, amylase clearances of 30−50 ml/min can be achieved during hemodiafiltration [2]. Thus, during the standard treatment time of 12 h/week, amylase clearances can be obtained during hemodiafiltration which equal or surpass the excretory function of the kidneys for this molecule with a molecular weight of 57,000 daltons.

However, during a standard hemodialysis treatment there was no change in amylase activity in the blood during a 4-month period (table I), indicating that either the dose of therapy applied was inadequate or that − as long as the balance of a substance used as marker is unknown − changes in blood concentration are inadequate indicators of renal replacement therapy.

Fortunately, it appears that there is only a minimal albumin loss by the membrane, if any. Correspondingly, from the course of total protein, albumin, pseudocholinesterase and cholesterol there was no indication for a protein loss during the observation period of 4 months, nor was there clinical evidence of a nephrotic syndrome. Unfortunately, using a standard dose of dialysis, excretion of smaller peptide hormones by the F 60 hemodiafilter is far from being normalized, as demonstrated for PTH fragments. A recent study [3], in which the less permeable polyacrylonitrile membrane (RP 6) was used, showed that during a single hemofiltration only clinically irrelevant amounts of small gastrointestinal peptides hormones are removed. Furthermore, the production of hormones, e.g. PTH, can be enhanced in renal failure, thus masking the removal of even relevant quantities of the hormone. As demonstrated for insulin, plasma concentrations of small gastrointestinal hormones are an unsuitable parameter for excretion studies, since the influence of stimuli for hormonal secretion, e.g. glucose, has by far greater importance than the effects resulting from hormonal elimination.

## Clinical Conclusions

Despite membrane properties of the F 60 which excel all other dialysis membranes concerning the imitation of the excretory kidney function, prescription of a *standard* dose of dialysis with the extremely permeable polysulfone membrane F 60 over 4 months revealed neither clinical and biochemical advantages nor disadvantages. However, taking into account that for the first time the whole molecular spectrum of substances to be removed by the kidneys can be eliminated by an artificial membrane, a *large* dose of renal replacement therapy with the F 60 might offer the chance to improve dialysis therapy.

Since hemodialysis has disadvantages in the removal of larger potential toxins and hemofiltration has disadvantages in the elimination of small uremic solutes, long-term hemodiafiltration, using the maximum potential of the F 60 membrane, would be a rational approach to test this hypothesis.

## References

1   Henderson, L.W.; Leypold, J.K.; Frigon, R.P.; Uyeji, S.N.; Alford, M.: Slow flow
    hemofiltration improves solute transport. 2nd Ann. Wkshop Int. Soc. Hemofiltration,
    Milano 1984.
2   Wizemann, V.; Techert, F.; Weber, N.; Brüning, S.: Single needle hemodialysis and
    hemofiltration in an extremely open membrane. Int. J. artif. Organs 6: 21–24 (1983).
3   Matthaei, D.; Ludwig-Köln, H.; Kramer, P.; Holzmann, H.; Morsches, E.; Benes, P.;
    Henning, H.; Klug, P.; Scheler, F.: Changes of plasma hormone levels in hemofiltra-
    tion. Int. J. artif. Organs 6: 21–24 (1983).

Prof. Dr. V. Wizemann, Zentrum für Innere Medizin, Universität Giessen,
Krankenhausstrasse 36, D-6300 Giessen (FRG)

Contr. Nephrol., vol. 46, pp. 69–74 (Karger, Basel 1985)

# Elimination of Low Molecular Weight Proteins with High Flux Membranes[1]

*A. Röckel, S. Abdelhamid, P. Fliegel, D. Walb*

Deutsche Klinik für Diagnostik, Wiesbaden, BRD

## Introduction

During the last decade, the important role of the kidney in low molecular weight (MW) plasma protein catabolism was established. Depending on their MW, configuration, and electrical charge, plasma proteins with a MW less than 65,000 daltons are filtered by the glomerulus, reabsorbed in the proximal tubule, and degraded within tubular cells. With increasing impairment of renal function, the plasma concentrations of most low MW proteins rise exponentially. In advanced renal failure, the plasma $\beta_2$-microglobulin, lysozyme, retinol-binding protein, and $\alpha_1$-antitrypsin concentrations are significantly increased.

These proteins are not eliminated with conventional Cuprophan (PT 150) membranes with a 400- to 3,000-dalton cutoff. In vitro studies have shown that cellulose acetate, polyacrylonitrile, polyamide, polysulfone, and modified Cuprophan membranes have a higher permeability and, therefore, should affect plasma concentrations of low MW proteins during hemofiltration, hemodiafiltration and hemodialysis.

To date, systematic in vivo investigations of low MW plasma protein removal in uremia are lacking. This study examines the elimination of some of these proteins during hemofiltration or hemodialysis with polysulfone (F 60), Cuprophan (Highflux), cellulose acetate (Duoflux), polyamide (FH 202), and polyacrylonitrile (PAN and RP 7 + 8) membranes.

[1] Supported by the Gesellschaft zur Förderung der Forschung (GFF) an der Deutschen Klinik für Diagnostik e.V.

## Methods

Studies were performed on 8 patients during routine hemodialysis or hemofiltration. Hemofiltration (HF) was done in the gravimetric, postdilution mode; exchange volume was 18 liters; mean fluid loss per treatment was 2 liters; the hemoprocessor (Sartorius) was used. For hemodialysis (HD) with the F 60 hemodiafilter the MTS A 2008 C (Fresenius) was used (3 h; mean fluid loss 2 liters).

The following high flux membranes were used: F 60 (Fresenius), Highflux, FH 202 (Gambro), Duoflux (Cordis Dow Corporation), RP 7 + 8 (Hospal) and PAN (Asahi). A 1:1 mixture of two solutions (Diaflac, SH: $Na^+$ 138 mval/l, $K^+$ 2.0 mval/l $Ca^{++}$ 4.2 mval/l, $Mg^{++}$ 1.5 mmol/l, $Cl^-$ 113.0 mmol/l, lactate 44.5 mmol/l, from Schi-wa, Glandorf, FRG) was used for substitution. The ultrafiltration rate was 75–85 ml/min, blood flow 320–450 ml/min and transmembrane pressure 150–170 mm Hg. F 60 was tested under ultrafiltration conditions (UFR 20 ml/min, blood flow 200 ml/min). For fistula access, siliconized fistula needles (diameter 2.0 mm; 14-gauge) from Bionic Medizintechnik (Bad Vilbel, FRG) and tube systems from Haemotronic (Mirandola, Italy) were used.

$\beta_2$-Microglobulin (11,800 daltons), lysozyme (15,000 daltons), retinol-binding protein (21,000 daltons), $\alpha_1$-antitrypsin (54,000 daltons), and albumin (66,500 daltons) served as marker substances. A radioimmunodiffusion technique [6] was applied for protein analysis using M or LC partigen plates (Behring, Marburg, FRG); a 1% agarose T plate (pH 8.6) from Behring was prepared for $\beta_2$-microglobulin analysis.

In order to evaluate membrane permeability, the mean sieving coefficient (SC) of each plasma protein was calculated using the concentration in blood at inlet ($C_{B_i}$), in blood at outlet ($C_{B_o}$), and in ultrafiltrate ($C_F$) according to the following formula:

$$SC = \frac{C_F}{(C_{B_i} + C_{B_o})/2} \; .$$

Table I. Sieving coefficients ($\frac{C \text{ ultrafiltrate}}{C \text{ plasma}}$) of plasma proteins before and after a 3-hour HD and HF, respectively, with high flux membranes ($^*p < 0.05$)

|  | $\beta_2$-Microglobulin | | Lysozyme | |
|---|---|---|---|---|
|  | 5 min | 180 min | 5 min | 180 min |
| F 60 | 0.65 ± 0.14 | 0.57 ± 0.13 | 0.49 ± 0.11 | 0.41 ± 0.06 |
| Highflux | 0.60 ± 0.11 | 0.59 ± 0.08 | 0.41 ± 0.02 | 0.37 ± 0.03 |
| Duoflux | 0.58 ± 0.03 | 0.51 ± 0.02 | 0.52 ± 0.05 | 0.51 ± 0.05 |
| FH 202 | 0.41 ± 0.07 | 0.20 ± 0.09* | 0.52 ± 0.09 | 0.50 ± 0.06 |
| Hospal 7 + 8 | 0.47 ± 0.03 | 0.40 ± 0.05 | 0.05 ± 0.02 | 0.03 ± 0.01 |
| PAN 15 | 0.008 ± 0.003 | 0.002 ± 0.001 | 0.01 ± 0.005 | – |
| PAN 20 | 0.01 ± 0.006 | 0.005 ± 0.002 | 0.008 ± 0.003 | – |

*Results* (table I)

*F 60*
Sieving coefficients were calculated for $\beta_2$-microglobulin (0.65), lysozyme (0.49), retinol-binding protein (0.06), $\alpha_1$-antitrypsin (0.04), and albumin (0.008).

During the 3 h hemodialysis period there was a significant decrease of the sieving coefficients of proteins >15,000 daltons ($p < 0.05$); sieving coefficients of $\beta_2$-microglobulin and lysozyme remained constant.

*Highflux*
Sieving coefficients were calculated for $\beta_2$-microglobulin (0.60), lysozyme (0.41), retinol-binding protein (0.06), $\alpha_1$-antitrypsin (0.03), and albumin (0.02).

There was no significant decrease of the sieving coefficients during hemofiltration.

*Duoflux*
Sieving coefficients were calculated for $\beta_2$-microglobulin (0.58), lysozyme (0.52), and retinol-binding protein (0.04). The membrane was relatively impermeable to proteins with a higher MW. During the hemofiltration, the $\beta_2$-microglobulin sieving coefficient decreased significantly.

| Retinol-binding protein | | $\alpha_1$-Antitrypsin | | Albumin | |
|---|---|---|---|---|---|
| 5 min | 180 min | 5 min | 180 min | 5 min | 180 min |
| $0.06 \pm 0.05$ | $0.003 \pm 0.001^*$ | $0.03 \pm 0.002$ | $0.003 \pm 0.001^*$ | $0.008 \pm 0.001$ | $0.001 \pm 0.0003^*$ |
| $0.06 \pm 0.01$ | $0.06 \pm 0.001$ | $0.03 \pm 0.002$ | $0.02 \pm 0.002$ | $0.02 \pm 0.003$ | $0.003 \pm 0.0001$ |
| $0.04 \pm 0.004$ | $0.04 \pm 0.01$ | $0.01 \pm 0.01$ | $0.01 \pm 0.01$ | – | – |
| $0.03 \pm 0.005$ | $0.01 \pm 0.002$ | – | – | – | – |
| – | – | – | – | – | – |
| – | – | – | – | – | – |

### FH 202

Sieving coefficients were calculated for $\beta_2$-microglobulin (0.41), and lysozyme (0.52). The membrane was impermeable to proteins with a higher MW. During hemofiltration the sieving coefficient of $\beta_2$-microglobulin decreased significantly.

### RP 7 and 8

The sieving coefficients of $\beta_2$-microglobulin and lysozyme were 0.47 and 0.05, respectively. The membrane was impermeable to proteins with a higher MW. During hemofiltration the sieving coefficient of $\beta_2$-microglobulin did not decrease significantly.

### PAN 15, PAN 20

Both membranes were impermeable to the tested plasma proteins. Low MW protein concentrations in the ultrafiltrate were too low to be detected.

## Discussion

In vitro membrane properties differ from in vivo results for a number of reasons. Protein layers change the permeability, especially for substances with a MW higher than 5,000 daltons [7, 8]. Under the described clinical conditions, F 60, Highflux, Duoflux, and FH 202 filters were permeable to molecules with a 15,000-dalton MW; $\beta_2$-microglobulin and lysozyme sieving coefficients were >0.4. The sieving coefficient of $\beta_2$-microglobulin decreased significantly with the FH 202 and Duoflux hemofilters during the 3-hour hemofiltration period. With the Highflux and the polysulfone membrane, $\beta_2$-microglobulin and lysozyme permeability were constant; the secondary membrane seems to be formed immediately after blood membrane contact.

The F 60 hemodiafilter was tested under conditions quite different from those used for other high flux membranes; ultrafiltration rate and transmembrane pressure were significantly lower, because this high flux filter would preferably be used under hemodiafiltration conditions. According to *Streicher and Schneider* [11], the sieving coefficients of $\beta_2$-microglobulin and lysozyme might increase using a higher ultrafiltration rate and transmembrane pressure. We therefore suggest that under comparable conditions this membrane eliminates molecules better in the higher molecular weight range than other high flux membranes. The cut-off of this polysulfone membrane seems to be similar to that of the peritoneal membrane [12].

One of the objectives of hemodialysis therapy is to avoid protein loss in patients with advanced renal failure, since this would worsen the frequently preexisting malnutrition. A long-term protein drain might be detrimental rather than beneficial, though there are 3- to 10-fold increases of low molecular weight protein concentrations in patients on regular dialysis treatment. Using these highly permeable high flux membranes, an albumin loss of 0.3−1.0 g/20 liters hemofiltrate is to be expected. Until now, however, no adverse reactions have occurred during a 6-month hemofiltration period using the Highflux from Gambro. The nutritional state (as determined by the plasma concentrations of short half-life proteins such as transferrin and pseudocholinesterase) was not adversely affected [13].

## References

1   Peterson, P.A.; Everin, P.E.; Berggard, I.: Differentiation of glomerular, tubular and normal proteinuria: determination of urinary excretion of beta-2-microglobulin, albumin and total protein. J. clin. Invest. *48:* 1189 (1969).
2   Bienenstock, J.; Poortman, J.: Renal clearance of 15 proteins in renal disease. J. Lab. clin. Med. *75:* 297 (1970).
3   Maack, T.H.: Renal handling of low molecular weight proteins. Am. J. Med. *58:* 57 (1975).
4   Kult, J.; Dragoun, G.P.: Low molecular serum proteins in uremic patients; in Heidland, Hennemann, Kult, Renal insufficiency, p. 115 (Thieme, Stuttgart 1975).
5   Kult, J.; Laemmlein, C.H.; Roeckel, A.; Heidland, A.: Beta-2-Mikroglobulin im Serum − Ein Parameter des Glomerulumfiltrates. Dt. med. Wschr. *19:* 1686 (1974).
6   Mancini, G.; Carbonara, A.O.; Heremans, J.F.: Immunochemical quantitations of antigens by single radial immunodiffusion. Immunochemistry *2:* 234 (1965).
7   Colton, C.K.; Henderson, L.W.; Ford, C.A.; Lysaght, M.: Kinetics of hemofiltration. J. Lab. clin. Med. *35:* 355 (1975).
8   Smeby, L.C.; Jörstad, S.; Wideröe, T.E.; Svartaas, T.M.: Transport of small, middle, and large molecular weight substances in a dual filtration artificial kidney. Middle molecules in uremia and other diseases. Artif. Organs *4:* suppl., p. 104 (1980).
9   Massry, S.G.; Goldstein, D.A.; Procei, W.R.; Kletzky, D.A.: Impotence in patients with uremia. A possible role for parathyroid hormone. Nephron *19:* 305 (1977).
10   Emmanouel, D.A.; Lindenheimer, M.D.; Katz, A.I.: Pathogenesis of endocrine abnormalities in uremia. Endocr. Rev. *1:* 28 (1980).

11  Streicher, E,; Schneider, H.: Polysulphone membrane mimicking human glomerular basement membrane. Lancet *II:* 382 (1982).
12  Roeckel, A.; Gilge, U.; Müller, R.; Heidland, A.: Elimination of low molecular weight proteins during hemofiltration and CAPD. Trans. Am. Soc. artif. internal Organs *28:* 382 (1982).
13  Roeckel, A.; Gilge, U.; Liewald, A.; Heidland, A.: Elimination of low molecular weight proteins during hemofiltration. Artif. Organs *6:* 307 (1982).

Priv.-Doz. Dr. A. Röckel, Deutsche Klinik für Diagnostik, Aukammallee 33, D-6200 Wiesbaden (FRG)

# Biocompatibility

Contr. Nephrol., vol. 46, pp. 75–82 (Karger, Basel 1985)

# Granulocyte Adherence Changes during Hemodialysis

*P. Aljama, A. Martin-Malo, R. Pérez, D. Castillo, A. Torres, F. Velasco*

Hospital 'Reina Sofía', University of Cordoba, Spain

## Introduction

Hemodialysis with Cuprophan membrane dialyzers is associated with a significant decrease in granulocyte count between 2 and 45 min after the onset of the dialysis session [1, 2]. However, when polyacrylonitrile membrane dialyzers are used no significant leukopenia is observed [3]. Moreover, the new synthetic non-Cuprophan membranes such as polysulfone or polymethylmethacrylate also induce milder granulocytopenia than Cuprophan. It has therefore been suggested that the degree of dialysis-induced leukopenia may well reflect the extent of the biocompatibility of the dialyzer membrane [4]. In fact, very recently *Ivanovich* et al. [5] reported that there is a good clinical correlation between dialysis leukopenia and frequency of symptoms of hemodialysis intolerance.

The mechanism by which this leukopenia appears remains open to discussion. *Craddock* et al. [6] postulated that it was due to intrapulmonary leukostasis resulting from the activation of the complement system with anaphylatoxin formation. Other investigators have challenged this hypothesis by demonstrating that leukopenia and complement activation can be dissociated when using certain types of membranes, in particular polycarbonate membranes [7].

Irrespective of the true mechanism of leukopenia, the rapid appearance and resolution during a few minutes in the dialysis session do indeed suggest that an intravascular margination of granulocytes underlies this phenomenon. Margination may be promoted by the decreased mobility [8] and increased adhesiveness [9] of granulocytes following exposure to the dialysis membrane. Consequently, *Guerrero* et al. [10] have demonstrated that dialysis granulocytopenia is associated with a striking increase in adherence during Cuprophan membrane hemodialysis and we have recently reported of dialysis membranes which do not induce significant leukopenia nor alter granulocyte adherence [4].

The purpose of the present work was to study whether or not several membranes have different effects on granulocyte adherence during dialysis and its relationship with the intensity of the induced leukopenia.

### Materials and Methods

9 chronic hemodialysis patients, who had been on maintenance dialysis for at least 6 months agreed to participate in the study. They were consecutively dialyzed with Cuprophan, ethylenevinylalcohol (EVAL), polyacrylonitrile (AN 69) and polysulfone (F 60) membrane dialysers. Blood was obtained from the arterial line before and 15 and 60 min after starting dialysis. White cell counts were performed with the model S Coulter counter. A differential count of 200 white cells was made for each sample.

Granulocyte adherence was measured with a modification of the assay proposed by *MacGregor* et al. [11]. Briefly, an uniform weight of nylon fiber which serves as the surface for adherence, is inserted into a 23 cm long Pasteur pipette producing an internal column of 1.5 cm in length. It is very important to accurately measure the weight and length of the column, the diameter of the lower end of the pipette and the volume of the sample to be tested, due to large variations which occur in adherence when these parameters are changed. The pre-column granulocyte count is compared with that in the effluent blood to calculate the percentage of granulocytes adhering to the nylon fiber. The greatest sensitivity of the assay system was achieved using 70 mg of nylon which led to a coefficient of variation for intra-test observations of 7.8%.

In 6 patients, we performed the following in vitro studies during dialysis with the 4 above-mentioned types of membranes. Whole blood was separated into cells and plasma by centrifugation. Predialysis cells were resuspended in plasma drawn at 15 min of hemodialysis, incubated for 15 min and tested for adherence changes in respect to the adherence value obtained before dialysis.

Each patient was his own control and granulocyte count and adherence were expressed as the percent change from predialysis measurement. Statistical analysis was undertaken with the Student's t test for paired observations.

### Results

In patients using Cuprophan, the mean ± SEM neutrophil count at 15 min had fallen significantly and was 24.9 ± 5.6% of the predialysis value (p < 0.001). By 60 min it had risen to 110.4 ± 6.9% (NS). The EVAL mem-

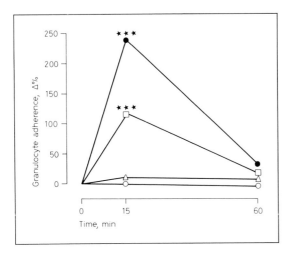

Fig. 1. In vivo granulocyte adherence changes during hemodialysis with different membranes: ● = Cuprophan; □ = ethylenevinylalcohol (EVAL); △ = polyacrylonitrile AN 69; ○ = polysulfone F 60. ***p < 0.001.

brane demonstrated a diminished response with neutrophil counts at 15 min of 60.2 ± 3.1% of predialysis measurement (p < 0.001). A statistical comparison of these results was significant (p < 0.005).

Neutropenia observed during hemodialysis using the synthetic membranes was negligible, being 88.1 ± 3.4% at 15 min with AN 69 and 86.3 ± 4.6% of predialysis value with F 60 (NS).

Granulocyte adherence in patients using Cuprophan membrane dialyzers increased at the same time as the neutropenia. In fact, at 15 min it reached 241.3 ± 11.6% of predialysis value (p < 0.001). During EVAL dialysis there was also a significant but less obvious increase in adherence of 118.1 ± 12.1% (p < 0.001). In both instances, granulocyte adhesiveness returned to the predialysis value at 60 min when the neutropenia recovered (fig. 1). However, AN 69 polyacrylonitrile dialyzers induced no significant increment in granulocyte adherence as it was 6.6 ± 2.4% of the predialysis value at 15 min. F 60 hemodialyzers showed comparable changes in adherence, being 5.1 ± 1.9% of the predialysis level (NS). It appears therefore that when there is no increase in granulocyte adherence, neutropenia does not occur.

Granulocyte adherence increased by 234.7 ± 14.2% of the predialysis level when predialysis cells were resuspended in plasma taken at 15 min of

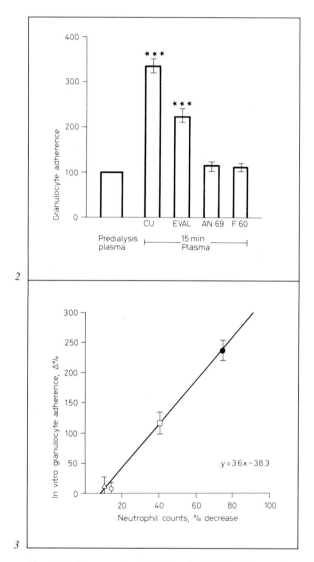

*Fig. 2.* In vitro granulocyte adherence with the 4 types of membranes studied: CU = Cuprophan; EVAL = ethylenevinylalcohol; AN 69 = polyacrylonitrile; F 60 = polysulfone. Predialysis adherence is considered to be 100% ***p < 0.001.

*Fig. 3.* Correlation between the ability of 15 min hemodialysis plasma to increased predialysis adherence of granulocytes in vitro with the degree of dialysis-induces neutropenia. ● = Cuprophan; □ = ethylenevinylalcohol; △ = polyacrylonitrile; ○ = polysulfone. The mean value at 15 min after the start of hemodialysis for both parameters is plotted (r = 0.996, p < 0.01).

dialysis from the patients using Cuprophan dialyzers. The same trend was observed with the EVAL membrane as predialysis white cells increased its adherence 120.4 ± 16.9% (p < 0.001) (fig. 2).

In contrast, there was no significant in vitro granulocyte adherence changes when these observatrions were repeated using AN 69 and F 60 dialyzers. As shown in figure 2, adherence increased only 6.1 ± 3.6% of predialysis measurement with AN 69 and 5.7 ± 3.5% with F 60 (NS).

A statistical comparison between the in vivo and in vitro changes in adherence induced by the different membranes used did not demonstrate significant differences. Therefore, it can be stated that the in vitro granulocyte adherence, with the present model test system, reflects the degree of biocompatibility of the dialyzer membrane. In order to support this hypothesis, we performed a linear regression analysis between the percent decrease in neutrophil counts and the mean increment in in vitro adherence changes with the 4 types of membranes studied. Despite the small number of events, a striking correlation was evident (r = 0.996, p < 0.01) as depicted in figure 3.

## Discussion

The present study shows, as have many others, that Cuprophan membranes produce a more marked neutropenia than their synthetic counterparts. It can be stated that the biocompatibility of EVAL dialyzers is midway between that of Cuprophan and AN 69 or F 60 dialyzers.

The long-term clinical consequences of these differences awaits further evaluation. In a previous study [3], we demonstrated that the white blood cell count in patients using AN 69 membranes was higher than in patients treated with Cuprophan dialyzers. One may speculate that this is probably important for transplantation because it may well influence tolerance to the immunosuppressive therapy. Nevertheless, comparison of the long-term health of patients treated with different kinds of membranes are clearly needed. In this regard there is a susceptible patient population suffering from the so-called 'dialysis-triggered asthma' [12, 13] whose dyspnea attacks are attributed to anemia, fluid overload, ischemic heart disease or uncontrolled hypertension; when they are switched to AN 69 dialyzers their dialysis-induced airway constriction ceases. This clinical problem may be the result of repeated pulmonary damage by intravascularly marginated granulocytes in the lungs during Cuprophan membrane hemodialysis [12].

In fact, the pulmonary vascular bed was found to be the major site of granulocyte sequestration, temporarily related to the granulocytopenia demonstrated in hemodialyzed dogs [14].

Furthermore, studying hemodialysis patients in a blinded prospective fashion, *Ivanovich* et al. [15] showed that intradialytic symptoms were statistically less severe with a lesser degree of neutropenia and granulocyte activation during hemodialysis. It seems therefore that biocompatibility might have some relevant clinical implications in relation to the dialysis-induced neutropenia. However, until very recently there have been only a few efforts to improve the blood compatibility of hemodialysis membranes. Such improvment could be achieved either by the development of new, more compatible materials or by modification of existing membranes.

Our present studies offer some additional information about the relationship between peripheral leukopenia and increase in granulocyte adherence. In fact, adherence reached the maximum at 15 min of dialysis and that coincided with the neutropenia nadir. If under experimental conditions adherence varies reciprocally with the neutrophil count [16], this parameter may reflect the intensity of granulocyte adhesiveness to the endothelial wall and hence the size of the marginated pool of granulocytes. *Craddock* et al. [17], in histologic sections of the lung of rabbits infused with Cuprophan-incubated plasma, documented the intravascular margination and leukostasis. We have failed to show these features in the lungs of dogs dialyzed with AN 69 membranes, whereas they were well demonstrated during Cuprophan membrane hemodialysis [unpubl. observations]. On the other hand, the adherence-augmenting activity was undetectable at 60 min after the start of the dialysis, when neutropenia recovered due to the release of the cell back to the peripheral circulation.

The in vitro experiments demonstrated that the factor responsible for the leukopenia, increased adhesivity and granulocyte margination is present in the plasma, dependent on the type of membrane used and activated by the contact of the plasma with the membrane. This is supported by the fact that there was a striking correlation between in vivo and in vitro adherence findings. In addition, these observations point away from direct cell-membrane interaction and reinforce the existence of a plasma factor that becomes activated with the contact during the firsts minutes of dialysis. Thus, the ability of 15 min dialysis plasma to increase granulocyte adherence, even in vitro, depends on the type of dialyzer used and is an important determinant of its biocompatibility. Although the determination of the nature of the plasma factor which promotes the increment in adherence during dialysis is beyond the scope of our present study, current work in our laboratory casts doubt on the central role of activated complement components. The present in vitro model test system is very useful not only for estimating the degree of biocompatibility of a given membrane but for studying in depth the properties and biological characteristics of this plasma factor.

In conclusion, we feel that the study of granulocyte adherence during dialysis reflects the biocompatibility of the dialyzer membrane. In this respect, the new synthetic membranes polyacrylonitrile AN 69 and polysulfone F 60 are shown to be superior to Cuprophan.

## References

1   Kaplow, L.S.; Goffinet, J.A.: Profound neutropenia during the early phase of hemodialysis. J. Am. med. Ass. 203: 1135-1137 (1968).

2   Gralt, T.; Schoroth, P.; De Palma, J.R.; Gordon, A.: Leucocyte dynamics with three types of hemodialysers. Trans. Am. Soc. artif. internal Organs 15: 45–49 (1969).

3   Aljama, P.; Conceicao, S.; Ward, M.K.; Feest, T.G.; Martin, H.; Graig, H.; Bird, P.A.E.; Sussman, M.; Kerr, D.N.S.: Comparison of three short dialysis schedules. Dial. Transplant. 7: 334–337 (1978).

4   Aljama, P.; Garin, M.; Torres, A.; Martin-Malo, A.; Moreno, E.; Perez-Calderon, R.: Haemodialysis leucopenia as an index of membrane biocompatibility. Contr. Nephrol., vol 37, pp. 129–133 (Karger, Basel 1984).

5   Invanovich, P.; Chenoweth, D.E.; Schmidt, R.; Klinkmann, H.; Boxer, L.A.; Jacob, H.S.; Hammerschmidt, D.E.: Cellulose acetate hemodialysis membranes are better tolerated than cuprophan. Contr. Nephrol., vol 37, pp. 78–82 (Karger, Basel 1984).

6   Craddock, P.R.; Fehr, J.; Dalmaso, A.P.; Brigham, K.L.; Jacob, H.S.: Hemodialysis leukopenia: pulmonary leukostasis resulting from complement activation by dialyser cellophane membranes. J. clin. Invest. 59: 879–888 (1977).

7   Aljama, P.; Bird, P.A.E.; Ward, M.K.; Feest, T.G.; Walker, W.; Tanboga, H.; Sussman, M.; Kerr, D.N.S.: Hemodialysis induced leucopenia and activation of complement: effects of different membranes. Proc. Eur. Dial. Transplant Ass. 15: 144–153 (1978).

8   Henderson, L.W.; Miller, M.E.; Hamilton, R.W.; Norman, M.E.: Hemodialysis leucopenia and polymorph random mobility: a possible correlation. J. Lab. clin. Med. 85: 191–197 (1975).

9   MacGregor, R.R.: Granulocyte adherence changes induced by hemodialysis, endotoxin, epinephrine and glucocorticoids. Ann. intern. Med. 86: 35–39 (1977).

10  Guerrero, I.C.; Schreiber, A.D.; MacGregor, R.R.: Studies of the plasma factor which augmented granulocyte adherence during hemodialysis. Nephron 27: 79–83 (1981).

11  MacGregor, R.R.; Spagnuolo, P.J.; Lentnek, A.L.: Inhibition of granulocyte adherence by ethanol, prednisone and aspirin, measured with a new assay system. New Engl. J. Med. 291: 642–646 (1974).

12  Aljama, P.; Brown, P.; Turner, P.; Ward, M.K.; Kerr, D.N.S.: Haemodialysis-triggered asthma. Br. med. J. *i:* 79–80 (1978).

13  Hanai, K.E.I.; Horiuchi, T.; Hanai, J.; Gotoh, H.; Hirasawa, Y.; Gejyo, F.; Aizawa, Y.: Hemodialysis-associated asthma in a renal failure patient. Nephron *25:* 247–248 (1979).

14  Toren, M.; Goffinet, J.A.; Kaplow, L.S.: Pulomonary bed sequestration of neutrophils during hemodialysis. Blood *36:* 337–340 (1970).

15  Invanovich, P.; Chenoweth, D.E.; Schmidt, R.; Klinkmann, H.; Boxer, L.A.; Jacob, H.S.; Hammerschmidt, D.E.: Symptoms and activation of granulocytes and complement with two dialysis membranes. Kidney int. *24:* 758–763 (1983).

16  Dale, D.C.; Fauci, A.S.; Wolff, S.M.: Alternate-day prednisone. Leukocyte kinetics and susceptibility to infections. New Engl. J. Med. *291:* 1154–1158 (1974).

17  Craddock, P.R.; Fehr, J.; Brigham, K.L.; Kronenberg, R.; Jacob, H.S.: Complement and granulocyte mediated pulmonary dysfunction in hemodialysis. New Engl. J. Med. *299:* 769–774 (1977).

Dr. Pedro Aljama, Servicio de Nefrología, Hospital 'Reina Sofía',
Avda. Menendez Pidal s/n, Córdoba (Spain)

Contr. Nephrol., vol. 46, pp. 83–91 (Karger, Basel 1985)

# Influence of High Permeability Synthetic Membranes on Gas Exchange and Lung Function during Hemodialysis

*S. Fawcett, N.A. Hoenich, C. Woffindin, M.K. Ward*

Department of Medicine, University of Newcastle upon Tyne, UK

## Introduction

During hemodialysis, a number of physiological and biochemical changes take place. The majority of these changes are of a therapeutic nature and benefit the patient. A number of the changes observed during treatment are at present the subject of increased research interest since their importance in the clinical management of the patient and their role in the morbidity associated with the procedure are yet to be fully appreciated. Three such aspects of hemodialysis are the transient neutropenia occurring during the first hour of treatment [1], the activation of the patient complement system by the blood membrane contact [2], and hypoxia or a fall in the arterial oxygen level ($PaO_2$) during treatment, a feature of hemodialysis that was first described in 1977 [3]. The magnitude of this fall is variable and ranges from 5 to 20% — and its onset, like that of neutropenia, is rapid but its nadir is delayed, unlike neutropenia return to predialysis levels are not restored until the termination of treatment.

Several mechanisms have been suggested to explain the cause of this hypoxia. *Craddock* et al. [2, 4] suggested that the decrease in oxygen tension was caused by leukocyte aggregation in the lungs secondary to complement activation by the membrane. More recent work confirms the importance of complement activation in causing neutropenia and sequestration of leukocytes in the lungs [5–9]. However, it remains uncertain whether this phenomena interferes with respiratory function. *Sherlock* et al. [10] proposed $CO_2$ loss into an acetate containing dialysate as being responsible while microemboli [11], increased oxygen consumption due to acetate metabolism [12], and pH changes induced by the treatment [13] have also been suggested as being responsible.

In the light of these suggestions, it is likely that hypoxia is multifactorial in origin. This study has been formulated to assess the role of membrane type, a factor known to influence leukopenia, by comparing the effects of Cuprophan® (Enka AG, Wuppertal, FRG) with synthetic high per meability membranes – polyacrylonitrile (PAN 15, Asahi Medical, Tokyo, Japan) and polysulfone F 60 (Fresenius AG, Bad Homburg, FRG) on hypoxia during hemodialysis. To study lung carbon monoxide diffusing capacity ($D_LCO$) when using the above membranes, since this parameter is an indicator of gaseous exchange across the alveolar membrane, and changes observed during hemodialysis would enable the establishment of an inter-relationship between hypoxia and neutropenia. The final objective of the study was to validate the hypothesis that pulmonary leukostasis mediates a fall in arterial oxygen tension.

## Materials and Methods

### Patients
14 patients whose age ranged from 32 to 62 years (mean 48.3 ± 9.8 years) receiving maintenance hemodialysis for chronic renal failure were studied. Ethical Committee approval and patient consent was obtained for all studies. None of the patients involved in the studies had signs of pulmonary insufficiency or congestive heart failure, although the absence of left ventricular hypertrophy was not included in the selection criteria.

### Dialysis Technique
Three groups of eight studies were performed using hemodialysers of comparable surface area: Fresenius/Dylade Hemoflow C 1.0 1.0 m² (Cuprophan), Fresenius Hemoflow F 60 1.25 m² (polysulfone) and Asahi PAN 15 1.2 m² (polyacrylonitrile) used with single patient proportionating systems in use at our center (Lucas II, Bellco Unimat and Fresenius/Dylade A 2008 E). All dialyzers were used with acetate-based dialysis fluid containing 40 mmol/l of acetate. Cuprophan and polyacrylonitrile containing dialyzers were used in the same patients, while those containing polysulfone membranes were used in the remainder of the group.

Circulatory access was via arteriovenous fistulae and blood flow was maintained at 200 ml/min throughout the studies.

All patients were systematically anticoagulated with heparin and treatment time ranged between 4 and 4½ h, with the fluid loss during treatments ranging from 1.5 to 2.5 kg.

### Blood and Expired Gas Measurement

Blood gas measurements were performed on 2-ml samples drawn from the arterial and venous segments of the extracorporeal circuit. Samples were drawn over a 30-second period into syringes whose dead space filled with heparin. Samples were drawn at the time of fistula needle insertion, as well as at 15, 60, 120 and 180 min during treatment, placed on ice and analyzed within 30 min of sampling by a Corning 168 pH meter.

Transfer factor ($D_LCO$) measurements were performed predialysis and again at times corresponding to the blood gas measurements with the patients in an upright position breathing room air through a Siebe-Gorman mouthpiece, prior to the measurements which were undertaken by the use of a Morgan Transfer factor machine. A single breath technique was used and patients were not permitted to smoke prior to these measurements. The inspired volumes were all greater than 1.2 liters.

### Leukocyte Count

Samples drawn predialysis as well as at 15, 30, 45, 60, 90, 120, 180 and 240 min during dialysis were analyzed using a Coulter counter for white blood cell counts.

### Results

### White Cell Counts

Changes in white cell count observed during treatment are summarised in figure 1. In patients using Cuprophan, the mean ($\pm$ SEM) white cell count by 15 min had fallen significantly and was 28.6 $\pm$ 1.8% of the predialysis value. By 30 min, it had risen by 46.5 $\pm$ 3.5% and continued to rise demonstrating rebound neutrophilia by the end of the first hour. The changes observed for polyacrylonitrile and polysulfone were less marked. In the former, the white cell count at 15 min had fallen to 73.5 $\pm$ 7.1%, while in the latter it fell to 78.9 $\pm$ 3.0%. By 30 min, the respective values were 78.5 $\pm$ 6.6% and 83.5 $\pm$ 3.5%.

Unlike Cuprophan no rebound neutrophilia was observed with either of the synthetic membranes.

A statistical analysis of the results showed the difference between Cuprophan and the synthetic membranes to be significant at the 0.1% level at the times ranging from 10 to 45 min after the commencement of dialysis. Although minor differences between polyacrylonitrile and polysulfone were noted, these differences failed to reach a statistical significance.

### $PaO_2$

Changes in $PaO_2$ (mm Hg) are shown in table I and the relative changes compared to predialysis levels in figure 2. When using Cuprophan, a rapid fall in the first 15 min of treatment was observed and the value at

60 min remained significantly below predialysis levels (p < 0.001). In contrast, with polyacrylonitrile PaO$_2$ was decreased only at 15 min, compared with predialysis levels. Polysulfone membrane failed to produce significant changes at any of the sampling times compared to predialysis values.

### PaCo$_2$

Arterial carbon dioxide levels (PaCo$_2$) expressed in mm Hg are shown in table II. The results obtained show no significant changes compared with predialysis values, although differences between the three groups were observed. These differences appear to be randomly distributed and no obvious correlation between membrane type and carbon dioxide loss emerges.

*Table I.* Arterial oxygen (PaO$_2$) changes during hemodialysis

| | Sampling time, min after start of dialysis | | | | | | |
|---|---|---|---|---|---|---|---|
| | predialysis | 15 | 30 | 60 | 120 | 180 | 240 |
| Cuprophan | 96.5 ± 1.4 | 84.7 ± 4.5 | – | 80.2 ± 3.6 | 82.5 ± 4.5 | 77.3 ± 2.6 | 78.2 ± 2.9 |
| | 100[1] | 87.7 ± 4.4 | | 83 ± 3.5 | 85.3 ± 4.1 | 80.1 ± 2.7 | 78.9 ± 3.2 |
| PAN 15 | 93.4 ± 4.0 | 83.1 ± 4.9 | – | 85.4 ± 6.1 | 85.1 ± 4.6 | 88.5 ± 6.4 | 83.9 ± 7.2 |
| | 100[1] | 88.9 ± 2.4 | | 95.5 ± 4.9 | 95 ± 3.7 | 94.5 ± 5.3 | 90 ± 7.5 |
| Polysulfone | 95.6 ± 3.3 | 97.4 ± 1.4 | 99.2 ± 3.1 | 93.6 ± 3.2 | 93.1 ± 4.5 | 93.7 ± 3.9 | 91.3 ± 3.6 |
| | 100[1] | 101.8 ± 1.4 | 103.7 ± 3.2 | 97.9 ± 3.3 | 97.3 ± 4.7 | 98.2 ± 4.1 | 95.5 ± 3.8 |

Results shown as mean ± SEM of eight observations (mm Hg).
[1] % change from predialysis.

*Table II.* Arterial carbon dioxide (PaCO$_2$) changes during hemodialysis

| | Sampling time, min after start of dialysis | | | | | | |
|---|---|---|---|---|---|---|---|
| | predialysis | 15 | 30 | 60 | 120 | 180 | 240 |
| Cuprophan | 34.4 ± 0.9 | 31.7 ± 1.0 | – | 33.6 ± 1.1 | 33.4 ± 1.1 | 34.2 ± 0.9 | 34.3 ± 0.9 |
| PAN 15 | 33.9 ± 1.2 | 33.1 ± 1.0 | – | 33.7 ± 1.2 | 32.9 ± 1.3 | 32.5 ± 1.1 | 31.7 ± 1.3 |
| Polysulfone | 34.8 ± 1.7 | 34.9 ± 1.6 | 32.9 ± 1.3 | 33.9 ± 1.1 | 34 ± 1.5 | 33.9 ± 1.2 | 32.6 ± 1.3 |

Results expressed in mm Hg as mean ± SEM of eight observations.

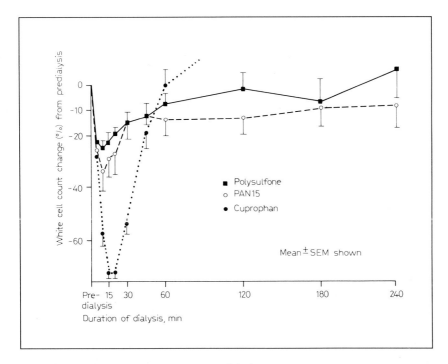

*Fig. 1.* White cell changes during hemodialysis.

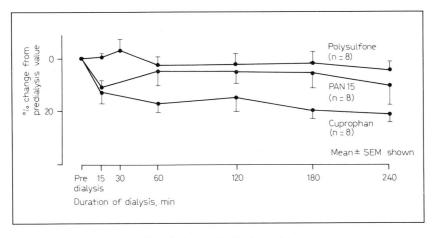

*Fig. 2.* Arterial oxygen (PaO$_2$) changes during hemodialysis.

*Table III.* $D_L CO$ changes during hemodialysis

| | Sampling time, min after start of dialysis | | | | | | |
| | predialysis | 15 | 30 | 60 | 120 | 180 | 240 |
|---|---|---|---|---|---|---|---|
| Cuprophan | 100 | $70.2 \pm 4.9$ | – | $61.3 \pm 4.3$ | $74.6 \pm 4.8$ | $78 \pm 6.1$ | $84.3 \pm 5.4$ |
| PAN 15 | 100 | $91.1 \pm 7.2$ | – | $86.5 \pm 6.4$ | $95.4 \pm 8.5$ | $92.3 \pm 7.6$ | $93.6 \pm 7.5$ |
| Polysulfone | 100 | $103 \pm 9.1$ | $100 \pm 5.3$ | $93 \pm 4.5$ | $92 \pm 6.1$ | $93 \pm 3.7$ | $95 \pm 6.1$ |

Results are shown as % change from predialysis $\pm$ SEM of eight readings.

*Transfer Factor ($D_L CO$)*

Transfer factor or carbon monoxide diffusing capacity expressed as a percentage change from predialysis values is shown in figure 3 and summarised in table III. With Cuprophan, a rapid fall in transfer factor was observed during the first hour of dialysis with the nadir of the fall occurring at 60 min, after which there was a gradual increase. Transfer factor remained below predialysis values even at 240 min, although the difference at this time was not significant statistically. Transfer factor during hemodialysis with polyacrylonitrile and polysulfone differed from this pattern in that the transfer factor with polyacrylonitrile also decreased but at no time was this decrease significant at the 5% level. Polysulfone failed to demonstrate significant differences from predialysis levels but remained below these levels from 30 min to the termination of treatment as for polyacrylonitrile.

*Discussion*

White blood cell counts performed confirm the differences described in the literature between cellulose based and synthetic membranes [5, 7, 14] and show the neutropenia induced by polysulfone to be comparable to other synthetic membranes.

Arterial oxygen tension ($PaO_2$) during hemodialysis is most severe with Cuprophan and is likely to be of clinical importance in patients with an already compromised cardiovascular status. The magnitude of observed hypoxemia appears to be related to the membrane type used and may be

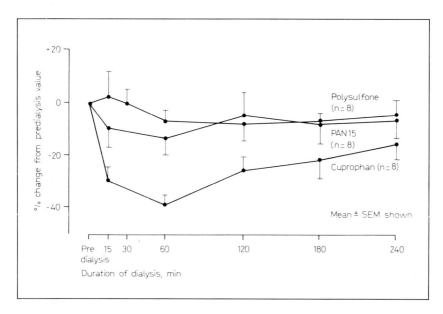

*Fig. 3.* $D_L CO$ changes during hemodialysis.

linked to the degree of neutropenia, although these observations cannot eliminate reflex hypoventilation being involved since dialyzer carbon dioxide loss occurs. This mechanism is not necessarily the dominant factor since Cuprophan used with bicarbonate containing dialysate still produces modest changes in $PaO_2$ [15].

Arterial carbon dioxide levels remained close to predialysis levels for all membranes.

The changes in transfer factor observed reflect both the changes in white cell count, as well as the decrease in $PaO_2$, with the most marked changes being noted for Cuprophan. However, the time course of these changes is not related to the neutropenia, which is reversed by the end of the first hour of treatment and may be due to acute damage to the lung during leukostasis.

Despite the similarity in neutropenia between polyacrylonitrile and polysulfone, changes in transfer factor appear delayed during polysulfone dialysis compared to the other groups.

Due to interdialytic weight gain in patients receiving regular dialysis treatment, there is a tendency for pulmonary edema to develop. As this fluid is removed during treatment, changes in lung function are likely and this may explain the gradual rise in transfer factor observed.

The long-term clinical implications of these phenomena are at present uncertain, but recently, evidence has emerged that lung function of patients undergoing regular dialysis treatment for chronic renal failure is abnormal in two major respects. Firstly, they show restrictive ventilatory defects [16] and a general decrease in $D_L CO$ partially due to anemia [17] and partially to an increase in the resistance of the alveolar membrane to gaseous diffusion [16]. It is quite feasible that repeated pulmonary leukostasis could result in irreversible changes in the alveolar membrane and hence decrease the transfer factor, and, in consequence, the use of more biocompatible membranes may be desirable to alleviate these possible complications of regular dialysis treatment.

## References

1   Kaplow, L.S.; Goffinet, J.A.: Profound neutropenia during the early phase of hemodialysis. J. Am. med. Ass. *208:* 133 (1968).
2   Craddock, P.R.; Fehr, J.; Dalmasso, A.P.; Brigham, K.L.; Jacob, H.S.: Hemodialysis leukopenia. J. clin. Invest. *59:* 879 (1977).
3   Aurigemma, N.M.; Feldman, N.T.; Gottlieb, M.; Ingram, R.H.; Lazarus, J.M.; Lowrie, E.G.: Arterial Oxygenation during hemodialysis. New Engl. J. Med. *297:* 871 (1977).
4   Craddock, P.R.; Fehr, J.; Brigham, K.L.; Kronenberg, R.S.; Jacob, H.S.: Complement and leukocyte-mediated pulmonary dysfunction in hemodialysis. New Engl. J. Med. *296:* 769 (1977).
5   Amadori, A.; Candi, P.; Sasdelli, M.; Massai, G.; Favilla, S.; Passaleva, A.; Ricci, M.: Hemodialysis leukopenia and complement function with different dialyzers. Kidney int. *24:* 775 (1983).
6   Ivanovich, P.; Chenoweth, D.E.; Schmidt, R.; Klinkmann, H.; Boxer, L.A.; Jacob, H.S.; Hammerschmidt, D.E.: Symptoms and activation of granulocytes and complement with two dialysis membranes. Kidney int. *24:* 758 (1983).
7   Chenoweth, D.E.; Cheung, A.K.; Henderson, L.W.: Anaphylatoxin formation during hemodialysis: Effects of different dialyzer membranes. Kidney int. *24:* 764 (1983).
8   Chenoweth, D.E.; Cheung, A.K.; Ward, D.M.; Henderson, L.W.: Anaphylatoxin formation during hemodialysis. Comparison of new and re-used dialyzers. Kidney int. *24:* 770 (1983).
9   Vinuesa, S.G. de; Resano, M.; Luno, J.; Gonzalez, C.; Barril, G.; Junco, E.; Valderrabano, F.: Leucopenia, hypoxia and complement activation in haemodialysis, three unrelated phenomena. Proc. Eur. Dial. Transplant Ass. *19:* 159 (1982).
10  Sherlock, J.; Ledwith, J.; Letteri, J.: Hypoventilation and hypoxemia during hemodialysis: reflex response to removal of $CO_2$ across the dialyzer. Trans. Am. Soc. artif. internal Organs *23:* 406 (1977).

11 Bischel, M.D.; Scoles, B.G.; Mohler, J.G.: Evidence for Pulmonary Microemboliza-
tion during Hemodialysis. Chest *67:* 335 (1975).

12 Patterson, R.W.; Nissenson, A.R.; Miller, J.; Smith, R.T.; Narins, R.G.; Sullivan,
S.F.: Hypoxemia and pulmonary gas exchange during hemodialysis. J. appl. Physiol.
*50:* 259 (1981).

13 Burns, C.; Scheinhorn, D.J.: Hypoxemia during hemodialysis. Archs intern. Med. *142:*
1350 (1982).

14 Hakim, R.M.; Lowrie, E.G.: Hemodialysis-associated neutropenia and hypoxemia.
The effect of dialyzer membrane materials. Nephron *32:* 32 (1982).

15 Fawcett, S.: B. Med. Sci.; thesis. University of Newcastle upon Tyne (1983).

16 Lee, H.Y.; Stretton, T.B.; Barnes, A.M.: The lungs in renal failure. Thorax *30:* 46
(1975).

17 Wolf, A.; Kummer, F.: Renal failure and carbon monoxide diffusing capacity of the
lung. Wien. klin. Wschr. *91:* 189 (1979).

Dr. N.A. Hoenich, Department of Medicine, The Medical School,
University of Newcastle upon Tyne, Newcastle upon Tyne, NE2 4HH (UK)

Contr. Nephrol., vol. 46, pp. 92–101 (Karger, Basel 1985)

# Influence of Various Membranes on the Coagulation System during Dialysis

*U. Hildebrand, E. Quellhorst*

Nephrologic Centre Niedersachsen, Hann. Münden, FRG

## Introduction

During the passage of blood through dialysis membranes in an extracorporeal circuit, interactions occur between blood components and exogenous surfaces, such as leukopenia, hypoxemia, complement activation and changes in platelet factor [1, 2]. These phenomena differ often with the different type of membranes [3, 4].

We have carried out similar investigations in some of our dialysis patients using Cuprophan and polysulfone in hemodialysis (HD) and Cuprophan, polysulfone, and PMMA in hemodiafiltration (HDF). In addition to general changes, we examined the possible influences of different membranes on the blood coagulation system.

## Methods

We examined 9 patients with end-stage renal failure aged between 21 and 70 years. They had been on dialysis treatment for 34 up to 108 months. 6 of them were treated with HD three times a week and 3 of them with HDF 3 times a week.

The dialyzers or hemodiafilters used were
Cuprophan: SMAD 140 (SMAD, Lyon, France, 1.4 m$^2$) and Hemoflow D 6 (Fresenius AG, Bad Homburg, FRG, 2.1 m$^2$); polymethylmethacrylate (PMMA): Filtryzer B$_1$L (Toray, Japan, 2.1 m$^2$); polysulfone: Hemoflow F 60 (Fresenius AG, 1.25 m$^2$).

Our examinations were carried out in the form of an A–B–A study regarding both HD and HDF. With HD we used the Cuprophan membrane SMAD 140 (A), followed by the polysulfone membrane F 60 (B) and then again the SMAD 140 (A). With HDF we compared the Cuprophan membrane D 6 (A) with the polysulfone membrane F 60 (B) and after that we again used the D 6 (A). In a second HDF series, we used the polymethylmethacrylate membrane Filtryzer $B_1L$ (A), then the polysulfone membrane F 60 (B) and then again the Filtryzer $B_1L$ (A).

As dialysis machine, we used the A2008 C (Fresenius AG) for HD and the same machine for HDF adapted with the ABC device (automatic balancing device). Dialysis time was 4 h for HD and HDF. All patients received identical amounts of heparin initially and as continuous dose during the treatment. The dialysis concentrate used for HD and HDF was an acetate concentrate (HMÜ II, Fresenius AG. For HDF we used 9 litres of substitution solution per treatment.

The following examinations were carried out at the beginning of dialysis, after 10, 30, 60, 120 and 240 min (measuring with the RTG S 801 Fresenius AG): PTT, TT, thromboplastin time, fibrin monomers, plasminogen, antiplasmin, antithrombin III (AT III), heparin, factor VIII.

## Results

Figure 1a, b shows the results of our determinations of leukocytes, thrombocytes, hemoglobin (Hb), and hematocrit (Hct). The comparison between leukocytes and thrombocytes using Cuprophan shows the known leukocyte drop and the variations of thrombocytes. With polysulfone and PMMA this is not so marked. Hct and Hb increase as expected during dialysis due to weight loss.

TT, plasminogen, and antiplasmin are not contained in a figure; TT was prolonged up to more than 5 min in the majority of the cases due to heparin and was thus not evaluated. There were no pathological changes of plasminogen and antiplasmin with the different membranes and types of dialysis.

The fibrin monomere level (fig. 2a, b) is the result of an increased formation of thrombin. It was increased in all patients and all dialyzers did not cause further increases, however, in the course of dialysis. We therefore conclude that a significant activation of coagulation has not taken place. This is supported by the fibrinogen level which remained normal, proving that fibrinogen has not been massively used. However, the fibrin monomere levels increased above normal showing that there was an activity of thrombin formation, which is always higher in dialysis patients than in healthy subjects.

In all HD and HDF treatments the thromboplastin time changed due to heparinisation. At the end of dialysis, it had again reached the initial value. It has often been observed in dialysis patients that the thromboplastin time is decreased already at the beginning of dialysis.

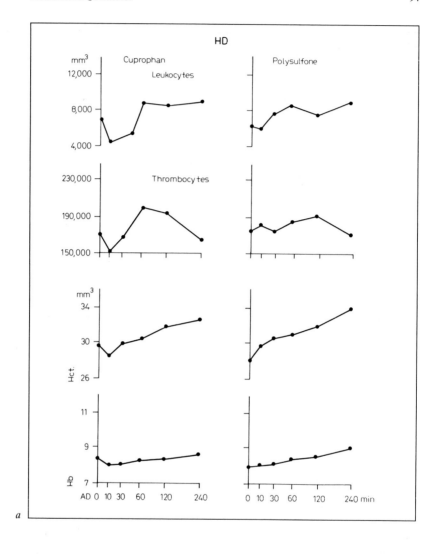

In order to obtain further results on blood coagulation using different membranes, we examined the blood by means of a resonance thrombograph (RTG). Modern resonance thrombography which is a further development of thrombelastography by *Hartert* [5, 6], has been used amongst other methods for examination of biocompatible membranes for the artificial heart. These were, however, always examinations with nonheparinized blood.

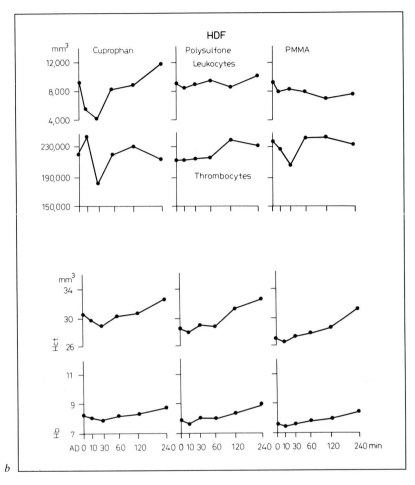

*Fig. 1.a, b* Leukocyte and thrombocyte counts and hematocrit and hemoglobin changes during HD *(a)* and HDF *(b)* with different membranes.

Figure 3 shows the scheme of a measuring curve of the RTG. Firstly, it contains the straight line (coagulation time). With the increase of the measuring curve the coagulation starts, which is practically the formation of a fibrin clot. It is measured with the fibrin time (f) and the fibrin amplitude (F). The decreasing part of the curve represents the platelet amplitude (p) which is the in vivo state of activity of thrombocytes and fibrinogen. In our case, the examinations were made more difficult as we used

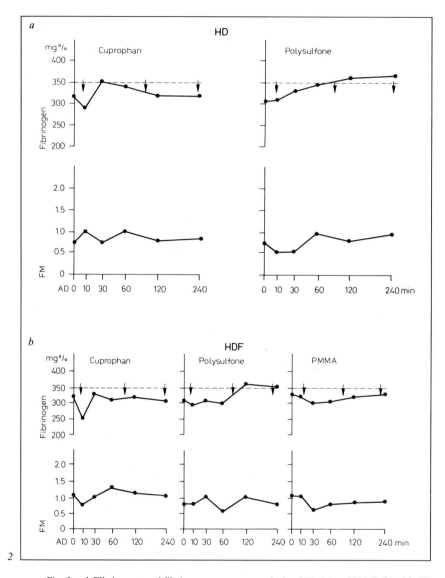

*Fig. 2.a, b* Fibrinogen and fibrinogen monomeres during HD *(a)* and HDF *(b)* with different membranes.

*Fig. 3.* Characteristics of the resonance thrombograph (RTG). r = Clotting time (min); f = fibrin time (min); F = fibrin amplitude (mm); P = platelet amplitude (mm) 5 min after maximum.

*Fig. 4.* Relationship between fibrin amplitude (F) and heparin consumption with four different dialysis membranes during 240 min of treatment; comparison of HD and HDF.

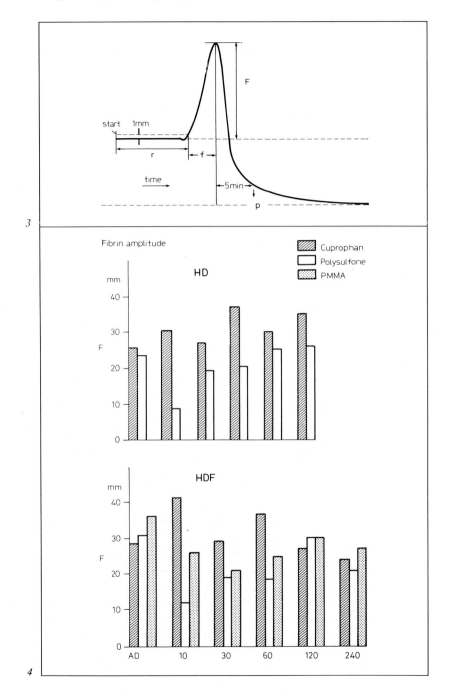

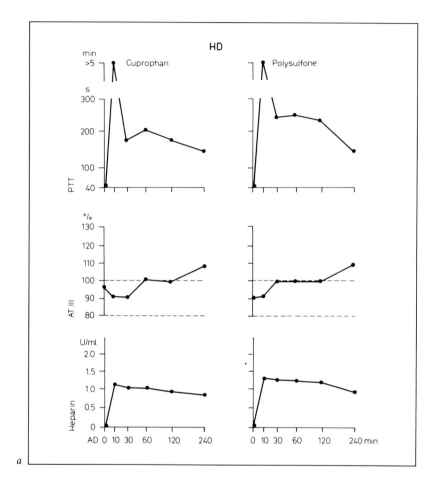

*a*

heparinized blood which influences the measuring figures. We therefore limited our evaluation of the F amplitude, the height of which showed the amount of heparin contained in blood after each specimen had been antagonized with 5 μl protamin. When all heparin was bound by antagonization, a long threaded clot was formed and created a high F amplitude. When the level of the biologically available heparin was too high so that not all heparin could be antagonized, we observed a short threaded clot with a small F amplitude or no increase of the curve at all.

According to the hypothesis that dialysis membranes require different quantities of heparin, membranes (Cuprophan) which require more hepa-

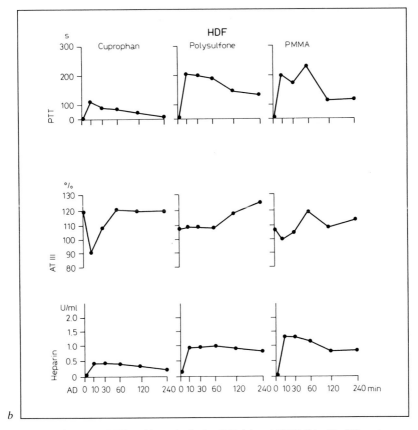

*b*

*Fig. 5.a, b* PTT, AT III and heparin during HD *(a)* and HDF *(b)* with different membranes.

rin during HD created high amplitudes, and membranes (polysulfone) which require less heparin created low amplitudes (fig. 4). We obtained similar results with HDF, where the height of the amplitudes for polysulfone and PMMA were similar (fig. 4). The different consumption manifestations especially at the amplitudes are very similar.

After having observed the tendency to higher heparin consumption with Cuprophan using the RTG, we observed similar findings with the control of partial PTT and measurable heparin in blood.

Figure 5a, b proves that after 30–120 min, PTT is higher with polysulfone than with Cuprophan membranes. The same applies to the heparin

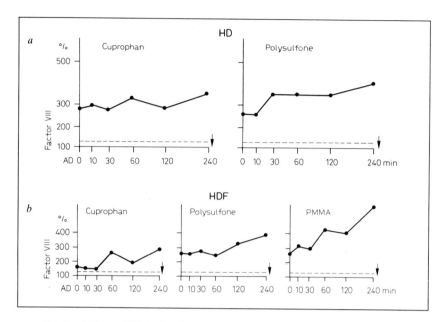

*Fig. 6.a, b* Factor VIII during HD *(a)* and HDF *(b)* with different membranes.

level in blood. It is by 10−25% higher with polysulfone and PMMA than with Cuprophan membranes and was highest with PMMA. At the end of treatment all membranes show almost equal levels of heparin, only with the Cuprophan membrane in HDF the heparin level remains at a low level.

It is known [7] that in dialysis patients factor VIII in blood is increased. We also observed that our patients had a factor VIII which was often twice as high as normal at the beginning of dialysis. During dialysis, factor VIII increased further (fig. 6a, b). With HDF the increase was more marked, and most significant with the PMMA membrane.

## Discussion

Our examinations show that during the first 2 h of treatment the consumption of heparin is higher using the Cuprophan membrane than with the polysulfone or PMMA membrane. This result has been found to be a tendency following RTG measurings but also after direct measurings of PTT and heparin blood levels.

The examinations show, furthermore, that all dialysis patients have increased fibrin monomere levels (above normal) as well as increased activity of factor VIII, independent of the membrane or dialysis method used.

Whereas the fibrin monomeres remain on the same level during all dialysis treatments, the activity of factor VIII continues to increase during dialysis. It has thus to be proven that further changes in the blood coagulation system prevail in HD in addition to normal anticoagulation with heparin which may increase hypercoagulability of uremic dialysis patients.

## References

1    Craddock, P.R.; Fehr, J.; Dalmasso, A.P.; Birgham, K.L.; Jacob, H.S.: Hemodialysis leukopenia. J. clin. Invest. *59:* 879 (1977).

2    Craddock, P.R.; Fehr, J.; Brigham, K.L.; Dronenbug, R.S.; Jacob, H.S.: Complement and leucocyte mediated pulmonary dysfunction in hemodialysis. New Engl. J. Med. *296:* 769 (1977).

3    Aljama, P.; Bird, P.A.E.; Ward, M.K.; Feest, T.G.; Walker, W.; Tanboga, H.; Sussmann, M.; Kerr, D.N.S.: Hemodialysis induced leukopenia and activation of complement: effects of different membranes. Proc. Eur. Dial. Transplant ass. *15:* 144 (1978).

4    Hakim, R.; Alfred, H.; Fox, K.; Corwin, B.; Lowric, E.: Biocompatibility of cellulosic and non-cellulosic hemodialysis membranes. Abstr. Ann. Meet. Natn. Kidney Foundation, 1979.

5    Hartert, H.: Blutgerinnungsstudien mit der Thrombelastographie, einem neuen Untersuchungsverfahren. Klin. Wsch. *26:* 577 (1948).

6    Harter, H.: Resonance-thrombography, theoretical and practical elements. Biorheology *18:* 693 (1981).

7    Schrader, J.; Kramer, P.; Scheler, F.: Blutgerinnungsveränderungen unter vier verschiedenen Dialyseverfahren. Nieren-Hochdruck-Krankh. *12:* 380 (1983).

Dr. med. U. Hildebrand, Nephrologisches Zentrum Niedersachsen, Vogelsang 37, D-3510 Hann. Münden 1 (FRG)

Contr. Nephrol., vol. 46, pp. 102–108 (Karger, Basel 1985)

# Complement Activation during Hemodialysis

## Comparison of Polysulfone and Cuprophan Membranes[1]

*S. Stannat, J. Bahlmann, D. Kiessling, K.-M. Koch, H. Deicher, H.H. Peter*[2]

Department of Medicine, Division of Nephrology and Division of Immunology, Medical School Hannover, FRG

## Introduction

The short-term biocompatibility of different dialysis and hemofiltration membranes used in extracorporeal blood circulations in renal failure patients has been investigated by several authors [1–4]. So far, mainly indirect parameters were studied, such as pulmonary function and leukocyte counts [1, 3, 4]. Only recently, quantitative changes in complement components and complement split products were measured [2]. Evidence was presented that cellophane membranes, in particular, cause temporary granulocytopenia and may be partly responsible for the hypoxemia observed during hemodialysis [2, 3].

With regard to the complement system, it is well known that synthetic materials can induce the alternative pathway of complement activation [5, 6]. Among the resulting split products, C3d has been shown to be the most stable one, and using a 'Double-decker- rocket immunelectrophoresis' (DDRIE) it can be quantified in EDTA plasma [7]. Furthermore, it has

[1] The investigation was partly supported by Deutsche Forschungsgemeinschaft, SFB 54 G3 and G13.

[2] The authors wish to thank staff of the dialysis unit at the Krankenhaus Oststadt/Hannover, as well as Mrs. *Berkefeld* and Mrs. *Kemper* for their technical assistance.

been shown that the C3d plasma levels reliably indicate the degree of complement activation.

In this study, we measured the C3 split product C3d before and at various times during hemodialysis with polysulfone and Cuprophan membranes. The results were compared with the leukocyte counts.

## Materials and Methods

### Patients

9 patients gave their informed consent to participate in this study. All of them had been under chronic hemodialysis for 2–6 years with three dialysis sessions per week.

In a first series of experiments, C3d plasma levels were determined in 3 patients who were regularly dialyzed for 5 h on a Cuprophan membrane plate dialyzer. Blood samples were collected before and after hemodialysis. 2 months later, the investigation was repeated, including taking a blood sample 2 h after starting dialysis.

Following this pilot study, a prospective study was commenced in 6 hemodialysis patients who had been dialyzed with Cuprophan membranes for years. 3 weeks before the beginning of the study they were changed to polyacrylonitrile membranes – which are suppposed to have a better biocompatibility [2] – in order to provide a 'washout period' from Cuprophan dialysis.

On the first day of the study, the patients were dialyzed with a polysulfone membrane hollow fiber hemodiafilter ( F 60, Fresenius, FRG). The next dialysis was performed with a Cuprophan membrane plate dialyzer using the same dialysis machine, acetate dialysate, tubing sets, and heparinisation (8000 I.E. total).

### Blood Sampling

1 ml of blood was taken from the arterial line in commercially available EDTA-prepared syringes (Monovetten, Sarstedt) before starting dialysis, and at 15, 60, 120 and 240 min thereafter for measurement of C3d, leukocyte count and hematocrit. During the Cuprophan dialysis, an additional blood sample was taken after passage of the dialyzer (venous line) at 15 min. After filling the hematocrit tubes, all samples were immediately centrifuged and stored at +4°C. Measurement of leukocyte counts and hematocrit were performed within 30 min and those of C3d within 2 h. Leukocyte counts and hematocrit values were determined by standard techniques.

### C3d Determination

A modified DDRIE [7] was used to quantify C3d plasma concentrations. C3d differs from native C3 and other C3 split products in molecular weight, and C3d presents only the antigenic determinant 'D' [8, 9]. Therefore, it can be separated by electrophoretic mobility and immunoprecipitation using specific antibodies (Dakopatts, Kopenhagen). Following DDRIE, the precipitation 'rockets' can be stained, measured and compared with standard 'rockets' of known C3d concentrations (C3d-Standard, Behringwerke/Marburg, FRG). The day-to-day reproducibility of this method is indicated by a coefficient of variation of 11%. The C3d levels observed in 56 healthy control persons ranged from 0 to 6 g/l · $10^{-3}$ (mean: 1.5 g/l · $10^{-3}$) in our laboratory.

## Statistical Evaluation

The Wilcoxon matched-pairs signed-rank test and a multiple analysis of variance (for interpretation of the C3d time courses) were applied. The limit of significance was set at p < 0.05.

## Results

In the pilot study with 3 patients on regular Cuprophan hemodialysis, the C3d concentrations were all within the normal range before starting dialysis. At the end of the dialysis, C3d plasma levels were increased in all 3 patients 5- to 12-fold (fig. 1). The result was reproducible when the same patients were reexamined 2 months later.

In the subsequent prospective study, the kinetics of C3d concentrations, leukocytes and hematocrit were investigated during polysulfone and Cuprophan membrane dialysis in 6 patients. The mean values before starting and at 15, 60, 120 and 240 min following exposure to the two different membranes are given in figure 2. It can be seen that in this group of patients the C3d plasma levels were slightly elevated before starting hemodialysis. Nevertheless, a significant increase was noted from 15 min on following polysulfone and Cuprophan membrane hemodialysis. The elevation of C3d was more pronounced in Cuprophan dialysis as compared to polysulfone dialysis, although this difference was not significant at the 0.05 level (p = 0.052).

The C3d plasma concentrations were maximal at 240 min: 32.8 ± 3.66 g/l · $10^{-3}$ (mean ± SEM) for Cuprophan (starting value: 16.7 ± 1.84 g/l · $10^{-3}$), and for polysulfone membrane dialysis 19.5 ± 4.57 g/l · $10^{-3}$ (starting value: 12.2 ± 5.78 g/l · $10^{-3}$).

The leukocyte counts showed a significant fall to 29.6% of the starting value within 15 min after beginning of Cuprophan dialysis, whereas a much lower decrease to 86.7% of the initial value, becoming significant at 60 min, was observed in polysulfone dialysis. At 240 min, the leukocyte counts were restored to starting levels in both dialysis techniques, with a slight increase in leukocytes to 128% in Cuprophan dialysis.

In all patients the variability of the mean hematocrit was minimal for the duration of both hemodialysis procedures.

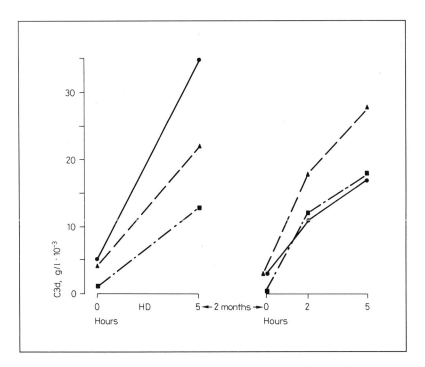

*Fig. 1.* C3d plasma levels in 3 patients before and after hemodialysis with Cuprophan membranes (left); data obtained after an interval of 2 months (right). Patients: ●——● = F.H.; ▲———▲ = G.K.; ■——■ = F.S.

## Discussion

Activation of complement by certain synthetic membranes is well documented [2, 3]. In the present study, Cuprophan hemodialysis membranes were shown to have a more pronounced effect on the activation of the complement component C3 than polysulfone membranes. The parameter used for the evaluation of C3 activation was the plasma level of the C3 split poduct C3d (fig. 3). It is known that C3d has a longer half-life than the anaphylatoxic split product C3a which has been shown to increase mainly during the first 15 min of dialysis concomitant to the drop of the leukocyte counts [2].

In contrast to C3a, the C3d plasma levels showed a delayed rise with a maximal increase in the first 60 min following initiation of hemodialysis,

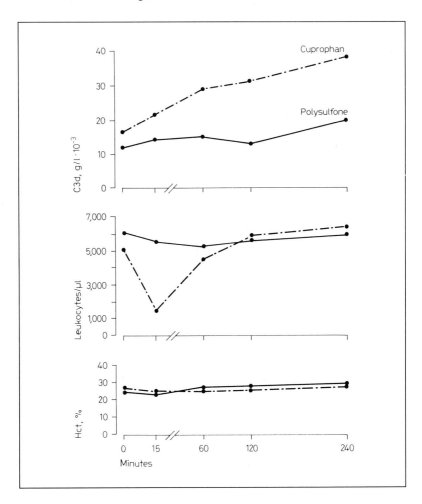

*Fig. 2.* Mean C3d concentrations, leukocyte counts and hematocrit values of 6 patients at various times during hemodialysis; comparison of dialysis with polysulfone (●——●) and Cuprophan membranes (●–·–·–●).

and a slower increase continuing up to 240 min, represented by a flatter slope of the C3d curve. Moreover, the kinetics of the C3d levels did not coincide with the drop of leukocyte counts, but they do parallel the reported accumulation of granulocyte elastase released from leukocytes during Cuprophan dialysis [10]. Thus, the sequence of events that occurs during hemodialysis on Cuprophan membranes may be as follows: (1) im-

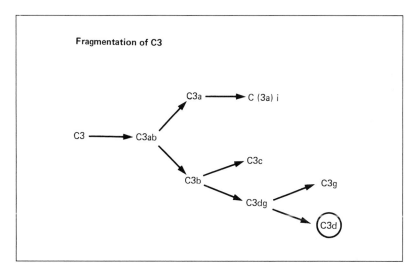

Fig. 3. Native C3 after activation (C3ab) and its cleavage into different fragments (C(3a)i: inactivated C3a).

mediate activation of C3 upon contact of blood with the membrane (see fig. 4); (2) liberation of the short-lived C3a which − besides for its vasoactive effects − is known to cause granulocyte aggregation and (3) gradual accumulation of the granulocyte elastase and C3d in blood, both delayed indicators of granulocyte and complement breakdown.

Although the different effects of hemodialysis with polysulfone or Cuprophan membranes on the activation of complement C3 has not been significantly proven yet, it is likely that this difference will become significant as more patients are examined and the polysulfone membrane is used without potting substance.

Apparently, polysulfone membranes seem to be biologically more compatible since they cause a lesser degree of acute C3 activation and a smaller drop in leukocyte counts. It should, however, be stressed that differences in long-term side effects from hemodialysis with Cuprophan or polysulfone membranes are far from being evident. Retrospective and prospective analysis of chronic hemodialysis patients will be needed to show the clinical relevance of our presented findings.

Nonetheless, it appears that complement activation measured by its breakdown product C3d may be a useful tool in judging 'biocompatibility' of synthetic materials used in extracorporeal blood circulation systems.

## References

1   Kaplow, L.S.; Goffinet, J.A.: Profound neutropenia during the early phase of hemodialysis. J. Am. med. Ass. *203:* 1135 (1968).

2   Chenoweth, D.E.; Cheung, A.K.; Henderson, L.: Anaphylatoxin formation during hemodialysis: Effects of different dialyzer membranes. Kidney int. *24:* 764 (1983).

3   Craddock, P.R.; Fehr, J.; Dalmasso, A.P.; et al.: Hemodialysis leucopenia. Pulmonary vascular leucostasis resulting from complement activation by dialyser cellophan membranes. J. clin. Invest. *59:* 879 (1977).

4   Ivanovich, P.; Chenoweth, D.E.; Schmidt, R.; et al.: Symptoms and activation of granulocytes and complement with two dialysis membranes. Kidney int. *24:* 758 (1983).

5   Harrison, R.A.; Lachmann, P.J.: The physiological breakdown of the third component of human complement. Molec. Immunol. *17:* 9 (1980).

6   Müller-Eberhard, H.J.; Schreiber, R.D.: Molecular biology and chemistry of the alternative pathway of complement. Adv. Immunol. *29:* 1 (1980).

7   Brandslund, I.; Sierstedt, H.C.; Svehag, S.-E.; Teisner, B.: Double-decker-rocket immunelectrophoresis for direct quantification of complement C3 split products with C3d specifities in plasma. J. immunol. Methods *44:* 63 (1981).

8   Sinosich, M.J.; Best, N.; Teisner, B.; Grudzinskas, J.G.: Demonstration of antigenic determinants specific for the split products of the 3rd complement factor, C3. J. immunol. Methods *51:* 355 (1982).

9   Teisner, B.; Brandslund, I.; Hau, J.; Svehag, S.-E.: Heterogenity in electrophoretic mobility of C3-derived molecules expressing D, but not C-epitopes following in vivo activation of the complement system. Acta pathol. microbiol. immunol. scand., C, Immunol. *91:* 85 (1983).

10  Hörl, W.H.; Jochum, M.; Heidland, A.: Release of granulocyte proteinases during hemodialysis. Am. J. Nephrol. *3:* 213 (1983).

cand. med. Sabine Stannat, Zentrum Innere Medizin, Abteilung Immunologie –
OE 6833 – Medizinische Hochschule Hannover, D-3000 Hannover 61 (FRG)

Contr. Nephrol., vol. 46, pp. 109–117 (Karger, Basel 1985)

# Release of Leukocyte Elastase during Hemodialysis
## Effect of Different Dialysis Membranes

*R.M. Schaefer[a], A. Heidland[a], W.H. Hörl[b]*

Departments of Medicine, Divisions of Nephrology, Universities of [a]Würzburg and [b]Freiburg, FRG

## Introduction

Polymorphonuclear granulocytes contain a broad variety of proteinases in order to repulse invading microorganisms [1]. These include the
neutral proteinases elastase [1, 2], cathepsin G [3] and collagenase [4], and
the two acid proteinases, cathepsin B and D [5]. Release of these proteinases occurs during cell death, phagocytosis, exposure to antigen-anti-
body complexes, activated complement components (C3a and C5a) and
toxic substances. In certain pathological situations massive release of
granulocyte proteinases may occur, resulting in tissue and plasma protein
degradation when the binding capacity of the plasma inhibitors is over-
whelmed.

During hemodialysis, an enhanced release of granulocyte elastase was
observed [6–9]. This phenomenon could be due to the contact of granulocytes with blood lines and the dialyzer membrane resulting in a so-called
'frustrated phagocytosis'. On the other hand, during hemodialysis complement activation occurs with the formation of C3a and C5a anaphylatoxins
inducing leukopenia and pulmonary vascular leukostasis [10, 11].
*Ivanovich* et al. [12] have demonstrated that cellulose acetate dialysis membranes were better tolerated than Cuprophan membranes, due to less complement activation. In addition, polyacrylonitrile, polycarbonate and
polyelectrolyte membranes have been reported to activate less complement than cellulose membranes [13–15]. The formation of C3a and C5a is

[1] The authors would like to thank Mrs. *M. Röder* for her excellent technical assistance
and Miss *M. Eiring* for her secretarial help.

only very modest during the early phase of dialysis using polyacrylonitrile membranes [16].

The present study was conducted to evaluate the effect of different hemodialysis membranes on: (1) the release of granulocyte elastase; (2) the concentration and activity of a $\alpha_1$-proteinase inhibitor, and (3) the 'unspecific' plasma proteolytic activity.

## Methods

### Patients

70 chronically uremic patients, aged $48.8 \pm 1.4$ years (mean $\pm$ SEM, range 20−69) undergoing regular hemodialysis treatment (RDT) for $39.4 \pm 3.5$ months (range 5−124) gave their informed consent to participate in the study. Hemodialysis was performed for 5 h three times per week with a glucose-free bath containing acetate. Cuprophan was used exclusively as the dialysis membrane. From this group 10 patients with high intradialytic granulocyte elastase-$\alpha_1$-proteinase inhibitor (E-$\alpha_1$PI) levels were chosen for determination of the effect of different dialysis membranes on the release of E-$\alpha_1$PI. Patients with diabetes mellitus or patients receiving corticosteroids or nonsteroidal anti-inflammatory agents were excluded from this study.

### Membranes

4 different membranes were selected for investigation. Cuprophan: CF 15−11 (Travenol, Deerfield/Ill.); polymethylmethacrylate: Filtryzer B2−150 (Toray, Tokyo, Japan); polyacrylonitrile: Biospal 3000S (Hospal, Lyon, France); polysulfone: Hemoflow F 60 (Fresenius, Oberursel, FRG).

### Sampling Procedures

For each membrane a run-in period of 3 weeks was chosen before the collection of blood was conducted. Whole blood samples were drawn from the patients' arteriovenous fistula prior to dialysis and 10, 30 and 60 min and at 60-min intervals thereafter up to 5 h. All blood samples were anticoagulated immediately with sodium citrate.

### Assay Procedure

Blood cells were counted by an electronic counter (Coulter-Counter Modell B). Plasma was separated from the sample within 30 min after its collection to prevent leakage of leukocyte constituents. The plasma specimens were stored at $-30°C$ until assayed. The measurement of plasma levels of the E-$\alpha_1$PI complex was performed with a highly sensitive enzyme-linked immunoassay [17]. The proteolytic activity of plasma samples was determined using azocasein as a substrate as previously described [18]. The inhibitory activities of $\alpha_1$-proteinase inhibitor ($\alpha_1$PI) and of $\alpha_2$-macroglobulin ($\alpha_2$M) were measured with a commercial test system (Boehringer, Mannheim, FRG). Plasma concentrations of $\alpha_1$PI and $\alpha_2$M were evaluated by a radial immunodiffusion plates (Behring-Werke AG, Marburg, FRG). Plasma total protein was measured by the method of Lowry et al. [19].

*Results*

Table I shows the increase of E-$\alpha_1$PI in 70 patients during hemodialysis according to the underlying kidney disease. Since the normal range of plasma E-$\alpha_1$PI is 60–90 ng/ml only patients with polycystic kidney degeneration had normal values before dialysis. During hemodialysis granulocytic elastase was released and plasma levels of E-$\alpha_1$PI were significantly enhanced in all groups of patients. Patients with diabetic nephropathy had the highest levels of E-$\alpha_1$PI before dialysis and the increase during HD was most pronounced. No sex difference in the release of granulocyte elastase during HD could be observed.

From this collective of 70 patients, 10 nondiabetic patients with a marked elevation of E-$\alpha_1$PI (600 ng/ml) after HD were chosen to study the effect of different dialysis membranes on the release of granulocyte elastase.

The group consisted of 5 men and 5 women, age 45.8 ± 3.9 years, regular dialysis treatment 42.4 ± 6.7 months, BUN 92.5 ± 1.2 mg/dl, creatinine 10.7 ± 0.1 mg/dl, hemoglobin 8.3 ± 0.2 g/l, hematocrit 27.1 ± 0.6%.

These patients were observed over a period of 5 months in order to determine the effect of different hemodialysis membranes. Blood samples were drawn from every patient following an adaptation period of 3 weeks for each dialyzer membrane. The effect of hemodialysis on white cell count

*Table I.* Effect of the underlying kidney disease on the release of granulocyte elastase during hemodialysis; mean values ± SEM of 70 patients; *p < 0.001

|  | Granulocyte elastase | |
|---|---|---|
|  | 0 h | 5 h |
| Kidney disease |  |  |
| Chronic glomerulonephritis (n = 45) | 132.4 ± 10.3 | 490.2 ± 40.2* |
| Chronic pyelonephritis (n = 9) | 138.9 ± 15.4 | 626.9 ± 71.6* |
| Polycystic kidney degeneration (n = 10) | 105.1 ± 13.5 | 487.7 ± 47.7* |
| Diabetic nephropathy (n = 6) | 189.0 ± 20.8 | 906.2 ± 69.0* |
| All patients (n = 70) | 133.2 ± 15.9 | 507.7 ± 47.9* |

*Table II.* Effect of different membranes on leukocyte count (cells/mm$^3$) during hemodialysis; data are expressed as mean values $\pm$ SEM of 10 patients; * p $<$ 0.001

| Membranes | Leukocyte counts | | | | |
|---|---|---|---|---|---|
| | 0 min | 10 min | 30 min | 60 min | 180 min |
| Cuprophan | 5,900 $\pm$ 600 | 1,900 $\pm$ 400* | 4,100 $\pm$ 600* | 5,400 $\pm$ 600 | 5,300 $\pm$ 500 |
| Polymethylmethacrylate | 7,400 $\pm$ 900 | 6,200 $\pm$ 1,000 | 5,800 $\pm$ 900 | 6,400 $\pm$ 900 | 6,500 $\pm$ 800 |
| Polyacrylonitrile | 5,000 $\pm$ 400 | 5,000 $\pm$ 300 | 5,200 $\pm$ 200 | 5,200 $\pm$ 300 | 7,100 $\pm$ 900 |
| Polysulfone | 6,800 $\pm$ 600 | 5,500 $\pm$ 600* | 5,600 $\pm$ 700 | 5,300 $\pm$ 500 | 5,700 $\pm$ 500 |

*Table III.* Effect of different dialysis membranes on plasma $\alpha_1$PI; mean values $\pm$ SEM of 10 patients; *p $<$ 0.05; **p $<$ 0.001

| Membranes | Concentration, mg/dl | | Activity, U/ml | | Increase of Hct, % |
|---|---|---|---|---|---|
| | 0 h | 5 h | 0 h | 5 h | % |
| Cuprophan | 212 $\pm$ 14 | 282 $\pm$ 17** (+33%) | 1.47 $\pm$ 0.10 | 1.69 $\pm$ 0.09* (+15%) | +10 |
| Polymethyl-methacrylate | 199 $\pm$ 17 | 245 $\pm$ 14** (+23%) | 1.54 $\pm$ 0.07 | 1.67 $\pm$ 0.07 (+8%) | +8 |
| Polyacrylonitrile | 202 $\pm$ 10 | 226 $\pm$ 8* (+12%) | 1.37 $\pm$ 0.14 | 1.86 $\pm$ 0.14** (+36%) | +13 |
| Polysulfone | 180 $\pm$ 17 | 193 $\pm$ 12* (+7%) | 1.67 $\pm$ 0.13 | 2.15 $\pm$ 0.19** (+29%) | +13 |

is shown in table II. With the Cuprophan membrane there was a marked decrease of leukocytes 10 min ($-68\%$; p $<$ 0.001) and 30 min ($-30\%$; p $<$ 0.001) after initiation of hemodialysis. Using polymethylmethacrylate ($-16\%$; p $<$ 0.001) or polysulfone ($-19\%$; p $<$ 0.001) membranes a smaller decrease of white cells could be observed. Patients on a polyacrylonitrile membrane showed no fall of leukocytes at all. The rise of plasma levels of the E-$\alpha_1$PI complex during hemodialysis is depicted in figure 1. Using Cuprophan membranes, the plasma E-$\alpha_1$PI reached a maximum level after 3 h of hemodialysis, whereas the polymethylmethacrylate-induced rise of E-$\alpha_1$PI reached a peak level after 5 h of hemodialysis (873 $\pm$ 94 ng/ml). The use of polyacrylonitrile and polysulfone membranes yielded much lower levels of E-$\alpha_1$PI during hemodialysis.

In table III, concentration and activity in plasma of $\alpha_1$-proteinase inhibitor ($\alpha_1$PI) are given according to the different membranes used. The

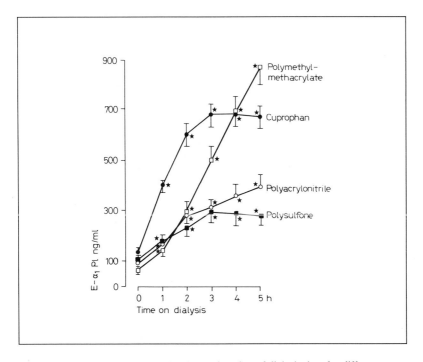

*Fig. 1.* Change of plasma E-$\alpha_1$PI levels as a function of dialysis time for different membranes. Data are expressed as mean ± SEM of 10 patients. *p < 0.001.

concentration of $\alpha_1$PI is markedly enhanced using Cuprophan (+33%; p < 0.001) or polymethylmethacrylate (+23%; p < 0.001) membranes. In contrast, hemodialysis with polyacrylonitrile or polysulfone membranes caused smaller increases of the concentration of the inhibitor, staying in the range of the elevation of hematocrit (Hct) during hemodialysis. On the other hand, there is only a small increase of the activity of $\alpha_1$PI using Cuprophan or polymethylmethacrylate membranes, whereas a pronounced elevation of the activity of this inhibitor is found using polyacrylonitrile (+36%; p < 0.001) or polysulfone (+29%; p < 0.001) membranes.

In spite of the release of granulocyte elastase during hemodialysis, a decrease of 'unspecific' proteolytic activity of plasma could be observed using azocasein as substrate. Figure 2 shows the effect of the different membrane material on the proteolytic activity of plasma. The smallest decrease occured with Cuprophan membranes (−34%; p < 0.001) whereas poly-

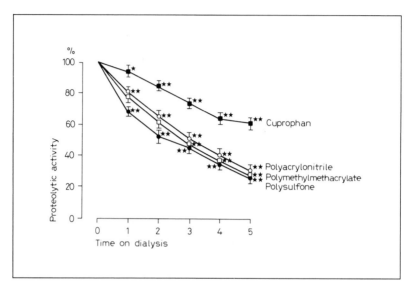

*Fig. 2.* Relative change of 'unspecific' plasma proteolytic activity as a function of dialysis time for different membranes. Data are expressed as mean < SEM of 10 patients. *p < 0.01; **p < 0.001.

acrylonitrile ($-68\%$; $p < 0.001$), polymethylmethacrylate ($-75\%$; $p < 0.001$) and polysulfone ($-75\%$; $p < 0.001$) membranes displayed marked decreases of unspecific proteolytic activity of plasma during hemodialysis.

## Discussion

In the present study, a pronounced decrease of leukocytes could be observed in the early phase of hemodialysis with Cuprophan membranes. It could be shown clearly that the fall of white cells depended strongly on the membrane used. Thus when using polymethylmethacrylate or polysulfone membranes, leukocytes decreased only to 84 and 81% of the pre-dialysis values. Polyacrylonitrile membranes induced no fall of white blood cells.

Plasma E-$\alpha_1$PI levels were found to be slightly increased in dialysis patients ($n = 70$). The group with polycystic kidney degeneration had the lowest levels and the diabetic hemodialysis patients had the highest pre-treatment levels of E-$\alpha_1$PI. The increase of E-$\alpha_1$PI during hemodialysis with Cuprophan membranes was similar in patients with chronic glomerulone-phritis, pyelonephritis and polycystic kidney degeneration, whereas diabet-ic hemodialysis patients displayed a more pronounced release of granulo-cytic elastase. Overall, the increase of E-$\alpha_1$PI did not correlate with the fall of white blood cells, which occurred in the early phase of hemodialysis, whereas the peak value of E-$\alpha_1$PI could be observed after 3 h of dialysis treatment using Cuprophan membranes. For evaluating the effect of differ-ent membranes on the release of granulocytic elastase, 10 nondiabetic pa-tients, having peak levels of more than 600 ng/ml, were selected. Again the release of granulocytic elastase strongly depended on the membrane used for dialysis. Cuprophan and polymethylmethacrylate membranes induced high levels ($680 \pm 80$ and $873 \pm 20$ ng/ml) of E-$\alpha_1$PI. In contrast, polyac-rylonitrile and polysulfone membranes induced moderate increases ($395 \pm 56$ ng/ml and $295 \pm 41$ ng/ml) of E-$\alpha_1$PI during hemodialysis.

The control of proteolytic activities in blood is exerted, primarily, by nine plasmatic proteinase inhibitors [20]. In vivo, 90% of granulocytic elas-tase is complexed by $\alpha_1$PI; therefore, concentrations and activities of this inhibitor were evaluated. During hemodialysis with Cuprophan or poly-methylmethacrylate membranes a marked increase of the concentration of $\alpha_1$PI ($+33\%$ or $+23\%$) occurred whereas the increase was only minor with polyacrylonitrile ($+12\%$) or polysulfone ($+7\%$), thus lying within the range of the increase of hematocrit. In contrast, the activity of $\alpha_1$PI was markedly increased using polyacrylonitrile ($+36\%$) or polysulfone mem-branes ($+29\%$), whereas Cuprophan and polymethylmethacrylate mem-branes induced only modest elevations ($+15\%$ and $+8\%$) of the $\alpha_1$PI activ-ity; thus suggesting that membranes with high release of granulocytic elas-tase induce high concentrations of $\alpha_1$PI in order to complex the proteinase elastase released from activated leukocytes. In contrast, membranes induc-ing only a moderate release of granulocytic elastase display only small in-creases of $\alpha_1$PI concentrations, whereas the activity of the inhibitor is mark-edly improved during hemodialysis. In order to determine 'unspecific' pro-teolytic activity in plasma the azocasein assay was performed. During dialysis with Cuprophan membranes the initially enhanced values were de-creased by 35%, whereas plasma proteolytic activity was reduced more than 60% using polymethylmethacrylate, polyacrylonitrile or polysulfone membranes.

The 'unspecific' proteolytic activity is generally elevated in chronic HD patients compared to healthy subjects. Until now, we have no explanation for this phenomenon, but we might speculate that low-molecular weight peptides, which are normally metabolized by the kidney, accumulate with the loss of kidney function. These peptides might display proteolytic activity.

Thus, using membrane materials with larger pores and higher 'cut-off' points, compared to Cuprophan, more proteolytic activity could be removed from plasma (fig. 2).

We would like to propose that parameters, such as release of leukocytic proteinases and elimination rates of proteolytic activity, should be included in the discussion about biocompatibility of hemodialysis membranes.

## References

1   Blondin, J.; Janoff, A.: The role of lysosomal elastase in the digestion of Escherichia coli proteins by human polymorphonuclear leukocytes. J. clin. Invest. *58:* 971 (1976).

2   Janoff, A.; Scherer, J.: Mediators of inflammation in leukocyte lysosomes. IX. Elastinolytic activity in granules of human polymorphonuclear leukocytes. J. exp. Med. *128:* 1137 (1968).

3   Rindler, R.; Schmalzl, F.; Braunsteiner, H.: Isolierung und Chrakterisierung der chymotrypsinähnlichen Protease aus neutrophilen Granulozyten des Menschen. Schweiz. med. Wschr. *104:* 132 (1974).

4   Ohlsson, K.; Olsson, J.: The neutral proteases of human granulocytes. Isolation and partial characterization of two graulocyte collagenases. Eur. J. Biochem. *36:* 473 (1973).

5   Baggiolini, M.; Bretz, U.; Dewald, B.: Subcellular localization of granulocyte enzymes; in Havemann, Jannoff, Neutral proteases of human polymorphonuclear leukocytes, pp. 3-17 (Urban, Schwarzenberg, Baltimore 1978).

6   Heidland, A.; Hörl, W.H.; Heller, N.; Heine, H.; Neumann, S.; Heidbreder, E.: Proteolytic enzymes and catabolism − enhanced release of granulocyte proteinases in uremic intoxication and during hemodialysis. Kidney int. *24:* (suppl. 16, p. 27, 1983).

7   Hörl, W.H.; Jochum, M.; Heidland, A.; Fritz, H.: Release of granulocyte proteinases during hemodialysis. Am. J. Nephrol. *3:* 213 (1983).

8   Heidland, A.; Hörl, W.H.; Heller, N.; Heine, H.; Neumann, S.; Schaefer, R.M.; Heidbreder, E.: Granulocyte lysosomal factors and plasma elastase in uremia: a potential factor of catabolism. Klin. Wschr. *62:* 218 (1984).

9   Hörl, W.H.; Heidland, A.: Evidence for the participation of granulocyte proteinases in intradialytic catabolism. Clin. Nephrol. *21:* 314 (1984).

10  Craddock, P.R.; Fehr, J.; Dalmasso, A.P.; Brigham, K.L.; Jacob, H.S.: Hemodialysis leukopenia: pulmonary vascular leukostasis resulting from complement activation by dialyzer cellophane membranes. J. clin. Invest. *59:* 879 (1977).

11  Craddock, P.R.; Fehr, J.; Brigham, K.L.; Kronenberg, R.S.; Jacob, H.S.: Complement and leukocyte-mediated pulmonary dysfunction in hemodialysis. New Engl. J. Med. *296:* 769 (1977).

12  Ivanovich, P.; Chenoweth, D.E.; Schmidt, R.; Klinkmann, H.; Boxer, L.A.; Jacob, H.S.; Hammerschmidt, D.E.: Symptoms and activation of granulocytes and complement with two dialysis membranes. Kidney int. *24:* 758 (1983).

13  Henderson, L.W.; Miller, N.E.; Hamilton, R.W.; Norman, N.E.: Hemodialysis, leukopenia, and polymorph random mobility − a possible correlation. J. Lab. clin. Med. *85:* 191 (1975).

14  Aljama, P.; Bird, P.A.E.; Ward, M.K.; Tauboger, H.; Sheridan, R.; Craig, H.; Kerr, D.N.S.: Hemodialysis induced leukopenia and activation of complement: Effect of different membranes. Proc. Eur. Dial. Transplant. Ass. *15:* 144 (1978).

15  Jacob, A.I.; Gavellis, G.; Zarco, R.; Perez, G.; Bourgoignie, J.: Leukopenia, hypoxemia and complement function with different hemodialysis membranes. Kidney int. *18:* 505 (1980).

16  Chenoweth, D.E.; Cheung, A.K.; Henderson, L.W.: Anaphylatoxin-formation during hemodialysis. Effect of different dialyzer membranes. Kidney Int. *24:* 764 (1983).

17  Neumann, S.; Henrich, N.; Gunzer, G.; Lang, H.: Enzyme-linked immunoasssay for human granulocyte elastase in complex with $\alpha_1$-proteinase inhibitor; in Hörl, Heidland, Proteases: potential role in health and disease, pp. 379−390 (Plenum Press, London 1984).

18  Hörl, W.H.; Schaefer, R.M.; Heidland, A.: Role of urinary alpha$_1$-antitrypsin in Padutin® (Kallikrein) inactivation Eur. J. clin. Pharmacol. *22:* 541 (1982).

19  Lowry, O.H.; Rosebrough, N.J.; Farr, A.L.; Randall, R.: Protein measurements with the Folin phenol reagent. J. biol. Chem. *193:* 265 (1951).

20  Travis, J.; Salvesen, G.S.: Human plasma proteinase inhibitors. A. Rev. Biochem. *52:* (1983).

Dr. R.M. Schaefer, Medizinische Universitätsklinik, Josef-Schneider-Strasse 2, D-8700 Würzburg (FRG)

Contr. Nephrol., vol. 46, pp. 118–126 (Karger, Basel 1985)

# Clinical Tolerance Test of a
# New Polysulfone Membrane

*P. Piazolo, W. Brech[1]*

Dialysis Center, Friedrichshafen, FRG

## Introduction

Whereas in the first 20 years of their use blood purification procedures were directed to the managing of uremic symptoms and the survival of the uremic patient, today good tolerance and patient comfort are requested as well.

To reach this high aim, it is necessary to individualise the dialysis regimen by adapting to the patient's time and space requirements (home dialysis, limited care dialysis, etc.), optimising the efficiency of dialyzers and dialysis machines (high-flux dialyzer, controlled ultrafiltration, etc.) and the application of the most suitable blood purification process (such as bicarbonate dialysis and hemofiltration).

The modification of the cellophane membrane and the innovation of synthetic membranes as the site of blood purification were involved in most improvements and had a decisive part in many cases. Only these permitted the realisation of alternative procedures such as hemofiltration and hemodiafiltration to counteract the intolerance reactions associated with hemodialysis. Not only did the use of cellulose acetate or polyacrylonitrile

[1] We are grateful to Mr. *Anger* and Mr. *Grözinger* for their assistance in performing the clearance determinations, to our dialysis staff for their patience, and to Mrs. *Aschermann* for the careful preparation of the manuscript.

membranes improve the elimination of solutes but resulted also in fewer intolerance reactions experienced by the patients. This innovative series also includes the polysulfone membrane employed in the Hemoflow F 60 capillary hemodiafilter (surface area 1.25 m$^2$). During the last 8 months, we investigated the performance of this dialyzer and particularly the clinical tolerance of patients.

## Patients and Methods

The study was performed under the best possible standardised conditions in 9 patients with end stage renal insufficiency (average age: 58 years) who had an endogenous creatinine clearance below 3 ml/min and had been on chronic hemodialysis for a minimum of 4 years and had been lately using a Cuprophan dialyzer (Travenol CF 15 11; 1.2 m$^2$). Generally, dialysis was performed for $3 \times 4$ h/week with the Cuprophan dialyzer as well as with the polysulfone hemodiafilter, using a Fresenius A 2008 C dialysis machine. We used water treated by reverse osmosis and dialysate with a sodium content of $135-143$ mval/l plus acetate. In addition to the blood tests which we do in all stable dialysis patients, we also performed in vivo clearance determinations for urea, creatinine, uric acid, phosphate and $\beta_2$-microglobulin (RIA, Pharmacia, Freiburg, FRG) during dialysis with the F 60 hemodiafilter, as well as blood counts and measurements of the serum $C_3$ complement fraction. The evaluation of the clinical tolerance of dialysis with the F 60 capillary was based on questioning of the patients, evalua-tion of individual dialysis protocols, information obtained from nursing staff and a physical examination of the patients.

The study included only patients being treated with the F 60 hemodiafilter for a minimum of 3 months (some up to 6 months). It also included patients with a marked dialysis discomfort syndrome of hypotension during dialysis, as well as 2 patients suffering from asth-matoid complaints when treated with a Cuprophan membrane. The medication (such as aluminium hydroxide, vitamin D substitution, and antihypertensive agents), diet and physical strain remained unchanged during the observation period.

## Results

Both in the low molecular as well as in the higher molecular range the F 60 capillary hemodiafilter showed an *efficiency* which to our knowledge has not been reached by any dialyzer with a comparable surface area (table I).

During the $3-6$ months of treatment with the F 60 hemodiafilter, the patients showed on average a lower predialysis blood phosphate level than with the Cuprophan dialyzer which was previously used for 1 year: urea (F

*Table I.* Plasma clearance in ml/min: urea, creatinine, uric acid, phosphate, $\beta_2$-microglobulin, F 60 hemodiafilter, blood flow 200 ml/min, dialysate flow 500 ml/min, mean TMP 50 mm Hg

| Date | Urea | Creatinine | Uric acid | Phosphate | $\beta_2$-Microglobulin |
|------|------|-----------|-----------|-----------|------------------------|
| 8/83 | $188 \pm 12$ | $170 \pm 8$ | $137 \pm 7$ | $147 \pm 16$ | $57 \pm 11$ |
| 2/84 | $194 \pm 6$ | $175 \pm 9$ | $136 \pm 11$ | $153 \pm 8$ | $59 \pm 6$ |

*Table II.* Behavior of leukocyte, thrombocyte and complement ($C_3$) levels before and after 10, 20, 30, 60 and 240 min of hemodialysis with the F 60 hemodiafilter

| Time | Leukocytes $\times 1/mm^3$ | Thrombocytes $\times 10/mm^3$ | $C_3$ complement mg/dl |
|------|---------------------------|-------------------------------|------------------------|
| Before dialysis | $7,626 \pm 1,360$ | $198 \pm 62$ | $73.0 \pm 17.2$ |
| After 10 min | $7,490 \pm 1,516$ | $178 \pm 53$ | $75.7 \pm 19.1$ |
| After 20 min | $7,518 \pm 1,268$ | $176 \pm 82$ | $70.3 \pm 18.5$ |
| After 30 min | $7,852 \pm 1,486$ | $181 \pm 71$ | $69.7 \pm 18.3$ |
| After 60 min | $7,829 \pm 1,875$ | $174 \pm 69$ | $71.4 \pm 17.3$ |
| After 240 min | $7,675 \pm 1,912$ | $178 \pm 58$ | $67.6 \pm 18.2$ |

60): $136 \pm 42$ mg%; urea (Travenol CF 15 11): $152 \pm 39$ mg%; phosphate (F 60): $4.2 \pm 1.8$ mg%; phospate (Travenol CF 15 11): $5.4 \pm 2.2$ mg%.

Predialysis serum creatinine, uric acid and electrolyte levels did not differ between Cuprophan and polysulfone dialysis.

The *biocompatibility* of the new dialysis membrane is evaluated against the changes occurring with Cuprophan dialysis: a sudden diminution of leukocytes with complement activation [2] in the early phase of hemodialysis, and at the same time a pulmonary dysfunction [3], usually taking an asymptomatic course. Our investigations showed no significant changes in leukocytes, thrombocytes and $C_3$ complement fraction levels in hemodialysis with polysulfone membranes (table II).

2 of the patients studied developed constant clinical symptoms of bronchial asthma with dialysis with a Cuprophan membrane and had to be treated with a polymethylmethacrylate dialyzer (Filtryzer) to remain symptom-free. Patient V.M. was symptom-free during and between F 60 dialysis, whereas patient P.D. showed neither subjective asthma complaints nor clinical signs of dyspnea or broncho-spasticity under

Table III. Symptoms of dialysis discomfort under Cuprophan and polysulfone membrane dialysis (per 100 treatment sessions)

| Symptom | Cuprophan | Polysulfone |
|---|---|---|
| Unrest | 45 | 14 |
| Excitability | 21 | 11 |
| Weariness | 31 | 13 |
| Drowsiness | 6 | 2 |
| Nausea | 18 | 9 |
| Vertigo | 9 | 8 |
| Headache | 14 | 10 |
| Itching | 34 | 9 |

hemodialysis but suffered from asthma attacks between dialyses. Her need for bronchospasmolytic agents (Berotec controlled dosage aerosol) was reduced to half of the quantity previously used between the dialysis.

A high rate of *eosinophilia* is found in hemodialysis patients using Cuprophan dialyzers: 27% [17], 25% [15], 13% [6]. *Michelson* et al. [14] attributed this to a reaction mediated by complement activation during contact with the Cuprophan membrane. Throughout years of treatment with a Cuprophan membrane 4 of our 9 patients now treated with polysulfone membranes had a constant eosinophilia with levels between 700 and 4,200 cells/mm$^3$. During the 3−6 months that they were treated with the F 60 dialyzer, the eosinophil count normalized to below 500/mm$^3$ in 3 patients and dropped constantly from above 3,200/mm$^3$ to below 1,000/mm$^3$ in patient P.D.

The dialysis *discomfort syndrome* increases in frequency and distinctiveness with the number of years on dialysis [8, 9]. Compared to Cuprophan membrane dialysis, the tolerance of the single dialysis improved clearly with the polysulfone dialysis. Especially the subjective complaints such as unrest, excitability, weariness, drowsiness, nausea, vertigo, headache and itching were reduced (table III). The interdialysis itching could not be definitively influenced during the brief observation period of the F 60 treatment. Patients with extensive itching and probably also marked hyperparathyroidism complained that this symptom continued also under polysulfone treatment.

The *objective symptoms* such as vomiting, muscle cramps and hypotension as well as circulatory shock and orthostatic reactions when rising from

*Table IV.* Incidence of objectifiable symptoms in hemodialysis with Cuprophan and polysulfone membrane dialyzers (per 100 treatment sessions)

| Symptom | Cuprophan | Polysulfone |
|---|---|---|
| Vomiting | 4 | 1 |
| Muscle cramps | 7 | 2 |
| Hypotension | 38 | 24 |
| Circulatory shock | 2 | 1 |
| Orthostatic reaction after dialysis | 8 | 2 |

*Table V.* Course of blood pressure and weight: administration of 10% saline solution, 0.9% saline infusion, 20% human albumin infusion and dextran infusion in haemodialysis with Cuprophan and polysulfone membrane dialyzers (average of 100 dialyses)

| | Cuprophan | Polysulfone |
|---|---|---|
| Predialysis blood pressure, mm Hg | 138.6/89.2 | 141.5/87.9 |
| Postdialysis blood pressure mm Hg | 110.1/85.4 | 122.3/80.1 |
| Weight loss, kg | 1.84 | 2.38 |
| 10% NaCl, ml | 34 | 15 |
| 0.9% NaCl, ml | 30 | 9 |
| 20% human albumin, ml | 8 | 3.4 |
| Dextran solution, ml | 12 | 0 |

the dialysis bed improved with the F 60 treatment (table IV). This improvement occurred in spite of the less favorable weight behavior under F 60 treatment.

The medication required for blood pressure regulation expresses a higher circulatory stability under polysulfone membrane dialysis (table V).

During the 3−6 months of the F 60 treatment, only slight variations in dry weight, physical powers (symptoms of dyspnea, heart trouble, feeling of weakness when performing familiar physical work) and the symptoms of well-being such as appetite, fluid intake, ability to enjoy food (such as meat), sexual activities, etc., were observed.

Generally the degree of anemia, need for blood transfusions, total serum protein, velocity of nerve conduction, skin color and frequency of hospitalisation remained unchanged.

*Table VI.* Clinical overall evaluation of the changes observed when transferring from Cuprophan to polysulfone membrane dialysis

| | |
|---|---|
| Subjective condition | in some cases markedly improved |
| Objectifiable symptoms | more stable circulation, less substitution of saline and plasma expanders less vomiting and muscle cramps |
| Measurable laboratory and long-term parameter | unchanged |

We observed two episodes of fever and chills, without any apparent site infection, during the F 60 dialysis. These episodes usually occur in Cuprophan dialysis as well. The blood cultures of the 2 patients were negative and pyrogens in the dialysate could not be detected.

Table VI gives an overall view of the clinical symptoms, showing the superiority of the F 60 treatment over the Cuprophan membrane dialysis, although an expression in figures is difficult.

*Discussion*

The results of our clinical study in a small group of stable hemodialysis patients and 2 patients with pulmonary dysfunction under treatment with a Cuprophan membrane point to a superiority of polysulfone membrane dialysis. This is substantiated by a better elimination of small and higher molecular substances which was already reported by *Streicher and Schneider* [19] in 1983. Furthermore, the polysulfone membrane has an excellant degree of biocompatibility which was so far only reached by non-Cuprophan dialyzers: polyacrylonitrile membrane [1], cellulose-acetate membrane [7] and polymethylmethacrylate membrane [16]. The patient treated with a polysulfone membrane experiences less dialysis discomfort and feels better between dialyses than a patient treated with a Cuprophan membrane. A similar superiority over the Cuprophan membrane was also demonstrated for another synthetic membrane − polyacrylonitrile [4].

Whilst the normal kidney achieves an endogenous creatinine clearance of approximately $90-120$ ml/min, a regular $1.2-1.5$ m$^2$ Cuprophan dialyzer will achieve an average creatinine clearance of approximately 11 ml/min (3 × 4 h/week of dialysis). Although under the same conditions, the polysulfone membrane will increase the creatinine clearance to 15 ml/min over the week, this higher efficiency does not express itself in a lower average predialysis serum creatinine level. The predialysis urea and phosphate levels are lower than with Cuprophan membrane dialysis. When relating this increase in efficiency in the elimination of small molecular toxins with the use of polysulfone membranes to the results of the National Cooperative Dialysis Study [11], a lower morbidity for dialysis patients with lower predialysis serum urea levels is to be expected for the polysulfone membrane. Apparently, the hemodialysis procedure with a Cuprophan membrane remains to be associated with intolerance reactions which were numerically documented by *Kjellstrand* [8] and correspond also to our results.

The hemofiltration procedure is said to be associated with a better tolerance of blood purification since here the solutes are separated across a membrane by convection at a molecular cut-off within the range of the glomerular membrane. Consequently, especially the elimination of low molecular substances is markedly lower than in hemodialysis [5, 18]. The lower incidence of dialysis discomfort and disequilibrium with nausea, vertigo, headache, muscle cramps, etc. encountered with hemofiltration was explained by a more gentle clearance of small molecules, despite the fact that the blood level of those small molecular toxins is substantially higher than in hemodialysis. The higher clearance of middle molecular substances in hemofiltration [12] seems to be of great importance for the explanation of the better tolerance. Dialysis with a polysulfone membrane offers a comparably good efficiency in the elimination of middle molecular substances, which explains the similar favorable tolerance which was frequently demonstrated for the polyacrylonitrile membrane [4].

In a certain way, the F 60 represents an ideal blood purification device combining the advantage of a high-efficiency hemodialyzer (diffusive solute transport) with those of a high-efficiency hemofilter (convective solute transport). Furthermore, dialysis with the F 60 is associated with fewer disadvantages than hemofiltration and hemodiafiltration since a sufficient detoxication can also be achieved at a low blood flow and higher hematocrit [10, 13].

We consider the high ultrafiltration rate of the F 60 hemodiafilter to be a disadvantage. It requires a very exact fluid balancing system also in the upper ranges which is at present provided by only a few dialysis machines.

Furthermore, we have observed two febrile episodes in the polysulfone membrane dialysis group within a relatively short observation period. Possibly at a low ultrafiltration during dialysis a backfiltration of dialysate occurs when the blood passes through the capillary, thus permitting the penetration of pyrogens through microleaks in the membrane, which we could not prove. We estimate that the ultrafiltration rate should therefore be at least 500 ml/h.

Undoubtedly, polysulfone membrane dialysis offers the advantage of improving the clinical symptoms, especially circulatory stability on dialysis, as well as a higher life quality due to a more generous handling of fluid and protein intake between the dialyses.

However, this dialysis method, which at the time being is still expensive, must prove that it will remain superior to the Cuprophan membrane dialyses in the years to come.

## References

1   Aljama, P.; Conceicao, S.; Ward, M.K.; Feest, T.G.; Martin, A.M.; Craig, H.; Bird, P.A.E.; Sussmann, M.; Kerr, D.N.S.: Comparison of three short dialysis schedules. Dial. Transplant 4: 334 (1978).
2   Craddock, P.R.; Fehr, J.; Dalmasso, A.P.; Brigham, K.L.; Jacob, H.S.: Hemodialysis leukopenia. Pulmonary vascular leukostasis resulting from complement activation by dialyzer cellophane membranes. J. clin. Invest. 59: 879 (1977).
3   Craddock, P.R.; Fehr, J.; Brigham, K.L.; Kronenberg, R.S.; Jacob, H.S.: Complement and leukocyte-mediated pulmonary dysfunction in hemodialysis. New Engl. J. Med. 296: 769 (1977).
4   Funck-Brentano, J.L.; Man, N.K.; Sausse, A.: The effects of dialysis with polyacrylonitrile membrane on neuropathy and middle molecule weight toxins. Kidney int. suppl. 2, p. 52 (1975).
5   Henderson, L.W.; Colton, C.K.; Ford, C.A.: Kinetics of hemofiltration. Clinical characterisation of a new blood cleansing modality. J. Lab. clin. Med. 85: 372 (1975).
6   Hoy, W.E.; Cestero, R.V.M.: Eosinophilia in maintenance hemodialysis patients. J. Dial. 3: 73 (1979).
7   Ivanovich, P.; Chenoweth, D.E.; Schmidt, R.; Klinkmann, H.; Boxer, L.A.; Jacob, H.S.; Hammerschmidt, D.E.: Symptoms and activation of granulocytes and complement with two dialysis membranes. Kidney int. 24: 758 (1983).
8   Kjellstrand, C.M.: Current problems in long-term hemodialysis. Dial. Transplant 9: 295 (1980).
9   Lazarus, J.M.: Complications in hemodialysis: an overview. Kidney int. 17: 783 (1980).
10  Leber, H.W.; Wizemann, V.; Goebeaud, G.; Rawer, P.; Schütterle, G.: Hemodiafiltration. A new alternative to hemofiltration and conventional hemodialysis. Artif. Organs 2: 150 (1978).

11  Lowrie, E.G.; Laird, N.M.; Parker, T.F.; Sargent, J.A.: Effect of the hemodialysis pre-
    scription on patient morbidity. New Engl. J. Med. *305:* 1176 (1981).
12  Man, N.K.; Granger, A.; Rondon-Nucete, M.; Zingraff, J.; Jungers, P.; Sausse, A.;
    Funck-Brentano, J.L.: One year follow-up of short dialysis with a membrane highly
    permeable to middle molecules. Proc. Eur. Dial. Transplant Ass. *10:* 236 (1973).
13  Mann, H.; Stiller, S.; Gürich, W.: Hämofiltration: Möglichkeit und Grenzen der An-
    wendung in der Klinik. Biomed. Techn. *24:* 167 (1979).
14  Michelson, E.A.; Cohen, L.; Dankner, R.E.; Kulczycki, A.: Eosinophilia and pulmo-
    nary dysfunction during Cuprophan hemodialysis. Kidney int. *24:* 246 (1983).
15  Montoliu, j.; Lopez-Pedret, J.; Andreu, L.; Revert, L.L.: Eosinophilia in patients un-
    dergoing dialysis. Br. med. J. *282:* 960 (1981).
16  Ota, K.; Okazawa, T.; Kumagaya, E.; Agishi, T.; Sugino, N.; Mitani, N.; Fujii, Y.;
    Kimura, M.; Tsukamoto, H.; Tanzawa, H.; Sakai, Y.: Polymethylmethacrylate capil-
    lary kidney highly permeable to middle molecules. Proc. Eur. Dial. Transplant. Ass.
    *12:* 559 (1975).
17  Scheuermann, E.H.; Fassbinder, W.; Frei, U.; Koch, K.M.; Baldamus, C.A.:
    Eosinophilia in hemodialysis. Contr. Nephrol., vol. 36, p. 133 (Karger, Basel 1983).
18  Streicher, E.; Schneider, H.; Mylius, U. v.: Theoretische und technische Grundlagen
    der Haemofiltration. Nieren-Hochdruck-Krankh. *1:* 9 (1978).
19  Streicher, E.; Schneider, H.: Stofftransport bei Hämodiafiltration. Nieren-Hochdruck-
    Krankh. *12:* 339 (1983).

Priv. Doz. Dr. P. Piazolo, Internistische Gemeinschaftspraxis, Dialyseinstitut,
Werastrasse 33, D-7990 Friedrichshafen (FRG)

# Hemodynamics

Contr. Nephrol., vol. 46, pp. 127–133 (Karger, Basel 1985)

## Hemodynamics during Hemodialysis with Dialyzers of High Hydraulic Permeability

*M. Schmidt, W. Schoeppe, C.A. Baldamus*

Department of Nephrology, University Hospital Frankfurt, FRG

### Introduction

Intratreatment hypotension is one major complication of hemodialysis [1–3]. Hypotension presents an even greater restriction on reducing treatment time than does the necessary removal of uremic solutes [4, 5]. During recent years, alternative treatment regimes have been introduced which under comparable conditions have shown an improved hemodynamic response to intratreatment volume removal. This effect was clinically [7, 8] and experimentally [5, 6, 9, 10, 14] proven for hemofiltration (HF) and could be established [11, 12] also for hemodiafiltration (HDF). Improved intratreatment vascular stability has recently also been claimed [13] for hemodialysis with hemodialyzers of high hydraulic permeability (HF-HD). To investigate possible hemodynamic differences between HD, HF, HDF and HF-HD, a comparative study was performed in the same patients under comparable and standardized conditions.

### Patients and Methods

14 (6 female, 8 male) stable HD patients of a mean age of 39 years (23–69 years) gave informed consent to be investigated. Patients were on a regular 3 times weekly 4–5 h acetate HD schedule for more than 2 months (mean, 58 months). They did not suffer from systemic diseases. Adequate echocardiographic readings could be recorded in all of them. 4 patients (1 F, 3 M) were hypertensive (> 150/90 mm Hg) and 2 patients (1 F, 1 M) gained high weights (> 3 kg) between treatments.

*Table I.* Results

| | HD | | HF-HD | | HDF | | HF | |
|---|---|---|---|---|---|---|---|---|
| Weight loss, kg | 2.1 ± 0.8 | | 2.1 ± 1.0 | | 2.0 ± 0.8 | | 1.9 ± 0.8 | |
| Net UF rate, ml/min | 8.8 ± 3.2 | | 8.8 ± 4.3 | | 61 ± 6.5 | | 119 ± 10.6 | |
| Urea clearance, ml/min | 121 ± 14 | | 125 ± 13 | | 131 ± 25 | | 124 ± 15 | |
| | pre | post | pre | post | pre | post | pre | post |
| MAP, mm Hg | 96 ± 13 | 90 ± 18* | 97 ± 16 | 97 ± 17* | 96 ± 13 | 99 ± 16* | 95 ± 13 | 100 ± 10* |
| LVIDd, cm | 5.2 ± .9 | 4.6 ± 0.9 | 5.3 ± 1.0 | 4.5 ± 0.9 | 5.2 ± 1.0 | 4.6 ± 0.8 | 5.2 ± 0.9 | 4.4 ± 1.0 |
| LVIDs, cm | 3.3 ± 0.7 | 2.9 ± 0.8 | 3.4 ± 0.8 | 2.8 ± 0.8 | 3.4 ± 0.8 | 2.8 ± 0.7 | 3.3 ± 0.7 | 2.6 ± 0.7 |
| FS, % | 37.5 ± 8.9 | 37.5 ± 8.2 | 35.0 ± 7.5 | 38.0 ± 8.7 | 33.8 ± 7.2 | 37.8 ± 9.0 | 35.8 ± 8.0 | 41.0 ± 8.1 |
| Vcf, cps | 1.2 ± 0.3 | 1.4 ± 0.4 | 1.1 ± 0.2 | 1.5 ± 0.4 | 1.1 ± 0.2 | 1.3 ± 0.4 | 1.1 ± 0.3 | 1.6 ± 0.4 |

* Significant intertreatment difference (HD – HF-HD, HDF, HF).

Patients were studied at weekly intervals on the same weekday and at the same time of day under standard conditions: constant treatment time, linear weight reduction, comparable weight at start of treatment and comparable weight loss within each patient. Small solute clearances during treatment were comparable in all treatment regimes.

HD and HF-HD were performed with an A 2008 C dialysis machine (Fresenius AG, Bad Homburg, FRG) controlling in- and outgoing dialysate volumetrically. For HD the 1.3 m² Cuprophan® (8 μm) capillary dialyzer, Hemoflow D 6 (Fresenius AG, Bad Homburg, FRG) with an ultrafiltration coefficient of 14.6 ml/(h × mm Hg) was used. For HF-HD the Hemoflow F 60 (Fresenius AG), a capillary dialyzer equipped with a polysulfone membrane of 1.25 m² surface, a wall thickness of 40 μm and an ultrafiltration factor of 40 ml/(h × mm Hg) was used. The same dialyzer was applied in HDF, but here the A 2008 C was equipped with an automatic balancing device (SM 40017, Sartorius GmbH, Göttingen, FRG) for ultra-filtration and replacement fluid. During HF, two F 60 devices in parallel were used. Here ultrafiltrate and replacement fluid were gravimetrically balanced including weight loss by the Haemoprocessor S 40020 (Sartorius GmbH, Göttingen, FRG).

Dialysate and replacement fluid for HF and HDF were of similar constitution: $Na^+$ 140 mEq/l, $K^+$ 2,4 mEq/l, acetate 35 mEq/l, glucose 195 mg/dl. $Ca^{++}$ in dialysate (3.5 mEq/l) was lower than in replacement fluid (4.3 mEq/l) because in regard to total plasma calcium a greater proportion is filtrable than dialyzable [15].

Treatment started with an extracorporeal recirculation period (ECC) of more than 20 min without solute or water flux across the membrane, followed by 240 min of treatment. At the end of ECC as well as at the end of the treatment period a two-dimensional echocardio-graphic investigation was performed (Hewlett Packard 77020A). Blood pressure (cuff method) and pulse rate (ECG) were recorded every 20 min. The outcome measures were blood pressure and heart rate and echocardiographic data [16]: end diastolic (LVIDd) and end systolic (LVIDs) left ventricular internal diameters. Fractional shortening (FS) was calcu-

$$\text{lated: FS} = \frac{\text{LVIDd} - \text{LVIDs}}{\text{LVIDd} \times \text{LVET}} \text{ and given as percent. Mean rate of circumferential fiber short-}$$

$$\text{ening was calculated as Vcf} = \frac{\text{LVIDd} - \text{LVIDs}}{\text{LVIDd} \times \text{LVET}} \text{ and given as circumferentials per second}$$

and corrected for 1.73 m² body surface. LVET is the left ventricular ejection time.

For statistical analysis the mean values ± SDs were calculated and the paired t test was applied. The limit of significance was set at a p value of less than 0.05.

## Results

Results are given in table I. The standardized experimental conditions were fulfilled: weight loss varied only insignificantly and urea clearance dif-fered only slightly between the treatment modes. Since clearances were limited by ultrafiltration rate during HF, the speed of the blood pump had to be turned down during the other treatment modes. Ultrafiltration rate was adjusted to approximately 50% of that in HF. Differences at end of treatment reached significance between HD and all other treatment regimes. LVIDd as well as LVIDs decreased significantly during all treatments due to weight removal and a consequent of intravascular hypovolemia. Left ventricular fractional shortening did not change during HD but increased significantly during all other treatment modalities. Mean velocity of circumferential fiber shortening increased significantly during all treatments.

## Discussion

Due to the very standardized conditions of this study, the influence of nontreatment specific factors is greatly limited since the conditions were met in regard to pretreatment weight, weight loss, linearity of weight loss, small solute removal and treatment time.

In stable hemodialysis patients, blood pressure fell during HD but was maintained or increased during all other treatment regimes. At the end of treatment, HD values were significantly different from those of other treatment modalities. This pattern is well described for HD and HF [9, 14]. It is due to the lack of an increase in vascular resistance during HD, whereas the same patient is able to increase it adequately during HF [9, 14]. Aside from benefical clinical reports [11], an increase in vascular resistance is also reported for HDF [12] but these data were derived from acute renal failure patients who were subject to all influences seen on intensive care wards.

The echocardiographic data during HD are mainly supported by data in the literature [17–20]: decrease in LVIDd and LVIDs, unchanged FS and increase in Vcf. One very thorough study [21], however, differs from this data. It found a significant increase in FS during regular HD with weight loss as well as during HD without weight loss, but a significant decrease during pure ultrafiltration without fluid replacement. These patients, however, in contrast to our patients, had abnormal values for LVIDd, LVIDs, FS and Vcf, characteristic for patients with mild congestive heart failure. The main question to be raised in view of our data is, why blood pressure is not maintained and FS is not increased during HD in contrast to HF, HDF and HF-HD. Although not statistically significant, there seem to exist gradual hemodynamic differences between treatment modes: HF, being the most favorable form of treatment, followed by HDF and HF-HD and then HD.

Of the many factors contributing to hemodynamic instability [1, 2, 14], some are unlikely to play a major role in our experimental setup because the different procedures were performed under comparable conditions. This would hold for acetate load [6, 14, 22, 23], $pO_2$ and $pCO_2$ [24−26] and sodium balance [27, 28]. Differences in membrane biocompatibility and in convective transport, however, could account for the differences found and have to be discussed.

Hemodynamic differences seen in this study could well be attributed to differences in membrane material because only during HD the less biocompatible Cuprophan® [29−31] was used, whereas for all other treatments a polysulfone hollow fiber was applied. As one example of better biocompatibility polysulfone does not have the effect of C3 complement activation and early neutropenia [32]. On the other hand, *Quellhorst* et al. [33] demonstrated that the well-established hemodynamic differences between HF and HD were maintained even when both procedures were performed with a polyacrylonitrile membrane, which also does not activate complement and does not cause neutropenia [31].

The gradual differences between the other three treatment modalities (HF, HDF, HF-HD) would remain unexplained by biocompatibility. However, they could be explained by differences in quantity of convective transport. During HF, only convective transport occurred. During HDF net ultrafiltration rate accounted for about 50% of that during HF. Although during HF-HD the net ultrafiltration rate was identical to that during conventional HD, ultrafiltration and backfiltration of dialysate into the blood compartment seems to take place within the F 60 hemodiafilter [34]. The amount of this filtration and backfiltration at zero net ultrafiltration rate is estimated under in vivo conditions to be in the range of 20 ml/min [35]. This significant amount of convective transport during HF-HD could then support the interpretation that hemodynamic stability is linked to convection. How convective transport might influence the hemodynamic response to volume removal during ESRD treatment remains to be investigated.

### References

1   Keshaviah, P.; Ilstrup, K.; Constantini, E.; Berkseth, E.; Shapiro, F.: The influence of ultrafiltration and diffusion on cardiovascular parameters. Trans. Am. Soc. artif. internal Organs *26:* 328 (1980).

2   Henderson, L.W.: Symptomatic hypotension during hemodialysis. Kidney int. *17:* 571 (1980).

3   Degoulet, J.; Reach, I.; Aime, F.; Berger, C.; Goupy, F.; Jacobs, C.; Rogas, R.; Leg-
    rain, M.: IVe rapport cumulatif. Epidemiologie des complications. J. Urol. Nephrol.
    *12:* 925 (1977).
4   Cambi, V.; Buzio, C.; Arisi, L.; Calderini, C.; David, S.; Manari, A.; Bono, F.;
    Zanelli, P.: Vascular stability and middle molecules removal in hypertonic haemodiafil-
    tration. Proc. Eur. Dial. Transplant. Ass. *18:* 681 (1981).
5   Shaldon, S.; Deschodt, G.; Beau, M.C.; Claret, G.; Mion, H.; Mion, C.: Vascular
    stability during high flux haemofiltration. Proc. Eur. Dial. Transplant. Ass. *16:* 695
    (1979).
6   Hampl, H.; Paeprer, H.; Unger, V.; Ryzlewicz, T.; Fischer, C.; Cambi, V.; Kessel, M.:
    Hemodynamic studies, acid-base status and osmolarity in different hemodialysis proce-
    dures. Artif. Organs *2:* 348 (1978).
7   Quellhorst, E.; Schünemann, B.; Doht, B.: Hemofiltration − a new method for treat-
    ment of chronic renal insufficiency. Trans. Am. Soc. artif. internal Organs *23:* 681
    (1977).
8   Baldamus, C.A.; Schoeppe, W.; Koch, K.M.: Comparison of haemodialysis and post-
    dilution haemofiltration on an unselected population. Proc. Eur. Dial. Transplant. Ass.
    *15:* 228 (1978).
9   Shaldon, S.; Beau, M.C.; Deschodt, G.; Ramperaz, P.; Mion, C.: Vascular stability
    during hemofiltration. Trans. Am. Soc. artif. internal Organs *26:* 391 (1980).
10  Chen, W.T.; Chaignon, M.; Omvik, P.; Tarazi, R.C.; Bravo, R.L.; Nahamoto, S.: He-
    modynamic studies in chronic hemodialysis patients with hemofiltration/ultrafiltration.
    Trans. Am. Soc. artif. internal Organs *24:* 682 (1978).
11  Wizemann, V.; Rawer, P.; Schütterle, G.: Ultrashort haemofiltration: long term
    efficiency and haemodynamic tolerance. Proc. Eur. Dial. Transplant. Ass. *19:* 175
    (1982).
12  Wizemann, V.; Sychla, M.; Leber, H.W.: Simultaneous hemofiltration/hemodialysis
    versus hemofiltration and hemodialysis: hemodynamic parameters. Proc. Eur. Soc.
    artif. Organs *7:* 143 (1980).
13  Chanard, J.; Brunois, J.P.; Melin, J.P.; Lavand, S.; Toupance, O.: Long-term results
    of dialysis therapy with a highly permeabe membrane. Artif. Organs *6:* 261 (1982).
14  Baldamus, C.A.; Ernst, W.; Frei, U.; Koch, K.M.: Sympathetic and hemodynamic re-
    sponse to volume removal during different forms of renal replacement therapy. Neph-
    ron *31:* 324 (1982).
15  Fuchs, C.; Brasche, M.; Donath-Wolfram, V.; Kubosch, J.; Quellhorst, E.; Scheler,
    F.: Dialysate calcium and plasma calcium fractions during and after hemodialysis. Klin.
    Wschr. *53:* 39 (1979).
16  Feigenbaum, H.: Echocardiography (Lea & Febiger, Philadelphia 1981).
17  Vaziri, N.D.; Prakash, R.: Echocardiographic evaluation of the effect of hemodialysis
    on cardiac size and function in patients with end-stage renal disease. Am. J. med. Sci.
    *278:* 201 (1979).
18  Chaignon, M.; Chen, W.T.; Tarazi, R.C.; Nakamoto, S.; Salcedo, E.: Acute effects of
    hemodialysis on echocardiography-determined cardiac performance. Am. Heart J. *103:*
    374 (1982).
19  Hanrath, P.; Schweizer, P.; Bleifeld, W.; Brass, H.; Mann, H.; Bauerdick, H.; Effert,
    S.: Änderungen des linksventrikulären Querdurchmessers und des Kontrak-
    tilitätsverhaltens bei Hämodialyse. Dt. med. Wschr. *101:* 655 (1976).

20   Hung, J.; Harris, P.J.; Uven, R.F.; Tiller, D.J.; Keller, D.T.: Uremic cardiomyopathy
     − effect of hemodialysis on left ventricular function in end-stage renal failure. New
     Engl. J. Med. *302:* 547 (1980).

21   Nixon, J.V.; Mitchel, J.H.; McPhaul, J.J.; Henrich, W.L.: Effect of hemodialysis on
     left ventricular function. J. clin. Invest. *71:* 377 (1983).

22   Van Stone, J.C.; Bauer, J.; Carey, J.: The effect of dialysate sodium concentration on
     body fluid distribution during hemodialysis. Trans. Am. Soc. artif. internal Organs *26:*
     383 (1980).

23   Kirkendol, P.L.; Devia, C.J.; Bower, J.D.; Holbert, R.D.: A comparison of the car-
     diovascular effects of sodium acetate, sodium bicarbonate and other potential sources
     of fixed base in hemodialysate solution. Trans. Am. Soc. artif. internal Organs *23:* 399
     (1977).

24   Suutarinen, T.: Cardiovascular response to changes in arterial carbon dioxide tension.
     Acta physiol. scand. *67:* suppl. 266, p. 1 (1966).

25   Gregory, G.A.; Egerll, E.I.; Smith, N.T.; Cullen, B.F.: The cardiovascular effects of
     carbon dioxide in man awake and during diethylether anesthesia. Anesthesiology *40:*
     301 (1974).

26   Burnum, J.F.; Hickam, J.B.; McIntosh, H.D.: The effect of hypocapnia on arterial
     blood pressure. Circulation *9:* 89 (1954).

27   Gotch, F.A.; Sargent, J.A.: Hemofiltration: an unnecessarily complex method to
     achieve hypotonic sodium removal and controlled ultrafiltration. Blood Purification *1:*
     9 (1983).

28   Shaldon, S.; Baldamus, C.A.; Koch, K.M.; Lysaght, M.J.: Of sodium, symptomatology
     and syllogism. Blood Purification *1:* 16 (1983).

29   Chenoweth, D.E.; Cheung, A.K.; Henderson, L.W.: Anaphylatoxin formation during
     hemodialysis: effects of different dialyzer membranes. Kidney int. *24:* 764 (1983).

30   Craddock, P.R.; Rehr, J.; Dalmasso, A.P.; Brigham, K.L.; Jacob, H.S.: Hemodialysis
     leukopenia: pulmonary vascular leucostasis resulting from complement activation by
     dialyzer cellophane membranes. J. clin. Invest. *59:* 879 (1977).

31   Amadori, A.; Candi, P.; Sasdelli, M.; Massai, G.; Favilla, S.; Passaleva, A.; Ricci, M.:
     Hemodialysis leukopenia and complement function with different dialyzers. Kidney int.
     *24:* 775 (1983).

32   Streicher, E.; Schneider, H.: Polysulfone membrane mimicking human glomerular
     basement membrane. Lancet 1983/II, 1136.

33   Quellhorst, E.; Schünemann, B.; Hildebrand, U.; Falda, Z.: Response of the vascular
     system to different modifications of hemofiltration and hemodialysis. Proc. Eur. Dial.
     Transplant. Ass. *17:* 197 (1980).

34   Schmidt, M.; Baldamus, C.A.; Schoeppe, W.: Characterization of solute and solvent
     kinetics in hemodialyzers with highly permeable membranes. Am. Soc. artif. internal
     Organs, Abstr. *13:* 55 (1984).

35   Schmidt, M.; Schoeppe, W.; Baldamus, C.A.: Backfiltration in hemodialyzers with
     highly permeable membranes. Blood Purification (in press, 1984).

M. Schmidt, Department of Nephrology, University Hospital Frankfurt,
D-6000 Frankfurt (FRG)

Contr. Nephrol., vol. 46, pp. 134–150 (Karger, Basel 1985)

# Hemodynamic Studies of Diffusive and Convective Procedures Using a Polysulfone Membrane

*H. Schneider, E. Liomin, E. Streicher*

Department of Nephrology and Hypertension, Katharinenhospital, Stuttgart, FRG

## Introduction

Questions concerning the circulatory stability during dialysis therapy have regained interest in recent years due to the shortening of the treatment time and the increasing number of older patients with a cardiovascular history. Numerous papers indicate that the pathogenesis of ultrafiltration intolerance and the so-called symptomatic hypotension observed independent of the extent of the actual weight loss, seems to be of a multifactorial nature [8, 10, 19, 21, 23, 27, 32, 40]. The symptoms encountered here stretch from a general postdialytic performance insufficiency, headache, vomiting, and states of confusion to complications caused by cardiac arrhythmias and threatening drops in blood pressure which can complicate the dialysis treatment markedly. Attempts to improve this situation took various research groups in different directions: controlled ultrafiltration with programmable machines was introduced, and for the filtrative procedures − hemofiltration, hemodiafiltration and sequential ultrafiltration with subsequent dialysis − this brought about an enhanced cardiovascular stability [13, 25, 47, 50, 58, 62] by increasing the peripheral resistance. The reintroduction of the physiological bicarbonate buffer, preceded by the development of proportioning systems, resulted in an improvement in the circulatory regulation and an enhancement of the vigilance of the patients receiving this treatment [22, 29, 43, 48, 49].

The interpretation of the occasionally quite contradictary results of hemodynamic studies is aggravated since on the one side only a few intraindividual comparisons of different therapeutical modes are available and on the other side different membrane materials were used, which due to their varying transportation kinetics, permit only a limited comparison.

It was the objective of this study to determine the cardio-circulatory behavior in hemodialysis, hemofiltration and hemodiafiltration both with acetate and bicarbonate buffers under identical conditions in intraindividual comparisons using a highly permeable polysulfone membrane.

## Methods

We studied 5 patients on chronic dialysis (3 females, 2 males, average age 54.3 years) with a right-ventricular indwelling catheter under bicarbonate hemodialysis (HDB), hemofiltration (HFB) and hemodiafiltration (HDFB) and acetate hemodialysis (HDA), hemofiltration (HFA) and hemodiafiltration (HDFA). From the onset of the treatment, cardiac output (Oxicon), pulmonal-arterial pressure (Hellige), heart rate, blood pressure, blood gas analyses and plasma catecholamine levels (Tritium-Catecholamine Kit Biosigma) were determined at hourly intervals. Stroke volume, average arterial blood pressure and peripheral resistance were calculated. The uniform duration of the treatment was 4 h with a continuous weight loss due to controlled ultrafiltration. The hyperhydration present at the onset of the treatment was comparable for the same procedures using different buffers; however, they differed between the patients. When making intraindividual comparisons, it was attempted to keep the differences in weight reduction as low as possible. All patients were digitalized, had a therapy-requiring hypertension with normotonic regulation by beta-blockers and vasodilators.

The composition of the dialysate and, respectively, the substitution solutions differed only in their buffer content which was maintained at the same level of 35 mval/l for acetate and bicarbonate. The polysulfone membrane F 60 (Fresenius, Bad Homburg, FRG) was used in all studies. The respective kinetic data is published elsewhere [57].

## Results

The initial hemodynamic situation was reproducible in all patients: 3 patients had a high cardiac output and a high cardiac output heart failure, 1 patient had threshold findings and 2 patients had normal cardiac parameters at the onset of the therapy.

With all acetate procedures the stroke volume and cardiac output decreased with a more or less marked increase in the heart rate. In HDA the peripheral resistance decreased regularly, except for 1 patient. No decrease in the peripheral resistance was observed with the filtrative procedures HFA and HDFA, but a clear increase was measured. An adequate reduction of the pressures in the pulmonary circulation was usually not observed, especially in patients with an initial cardiac output heart failure under acetate buffer (fig. 1−3).

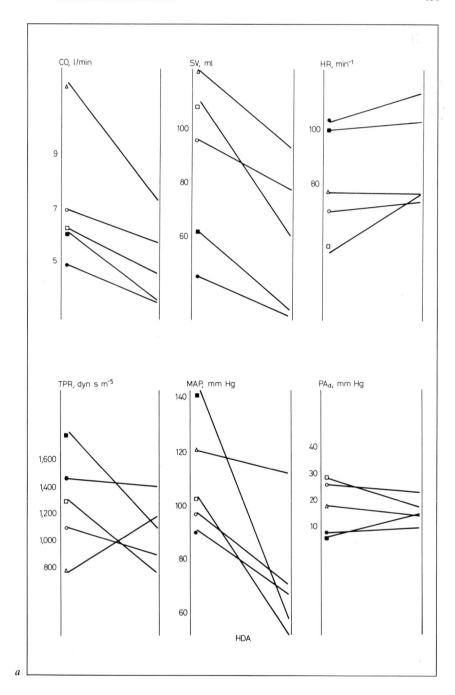

a

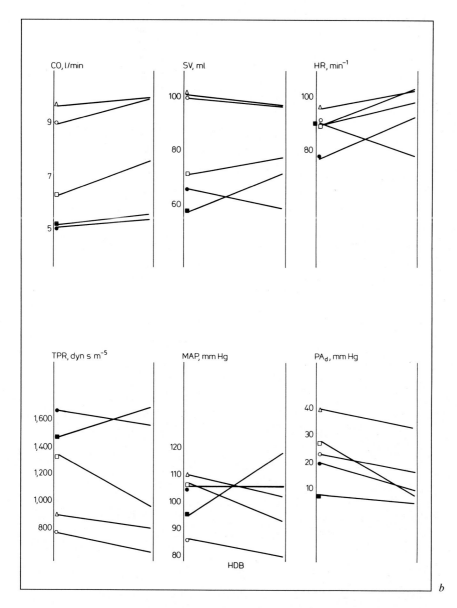

*Fig. 1.* Changes in hemodynamic parameters during 4 h of HDA (left) and HDB (right). △ = Patient I; ○ = patient II; □ = patient III; ■ = patient IV; ● = patient V. The following determinations were made at the onset and the end of the therapy: cardiac output (CO), stroke volume (SV), heart rate per minute (HR, min⁻¹), peripheral resistance (TPR), mean arterial pressure (MAP) and diastolic pulmonal-arterial pressure (PA$_d$).

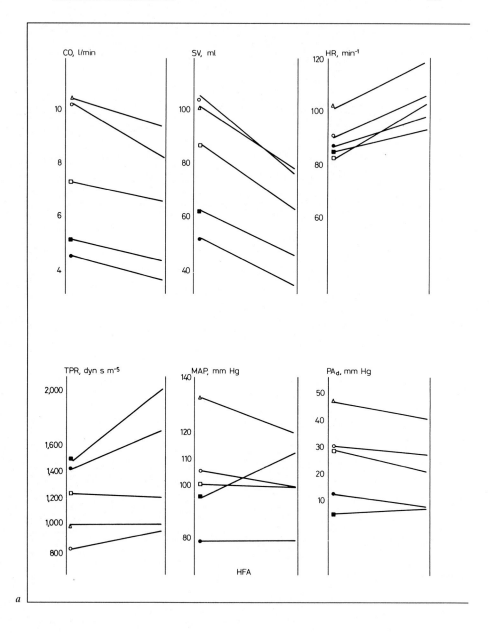

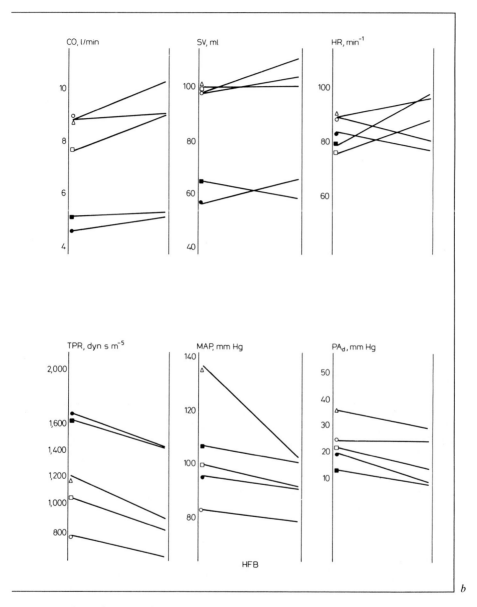

*Fig. 2.* Changes in hemodynamic parameters during 4 h of HFA (left) and HFB (right). The following determinations were made at the onset and the end of the therapy: cardiac output (CO), stroke volume (SV), heart rate per minute (HR, min⁻¹), peripheral resistance (TPR), mean arterial pressure (MAP) and diastolic pulmonal-arterial pressure (PA$_d$). Symbols as in figure 1.

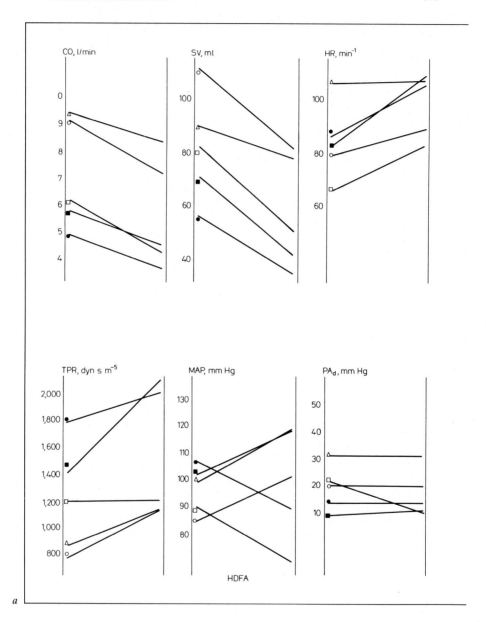

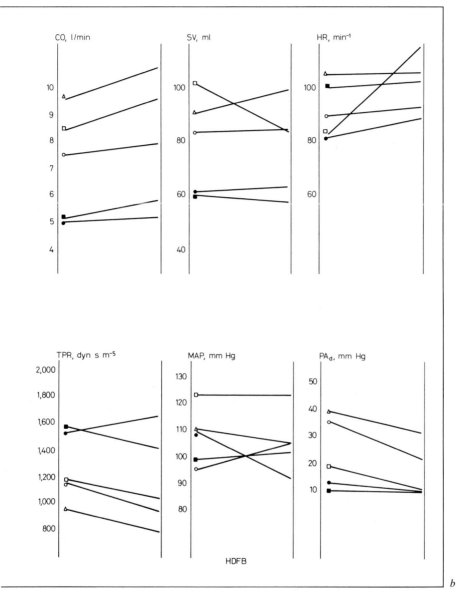

*Fig. 3.* Changes in hemodynamic parameters during 4 h of HDFA (left) and HDFB (right). △ = Patient I; ○ = patient II; □ = patient III; ■ = patient IV; ● = patient V. The following determinations were made at the onset and the end of the therapy: cardiac output (CO), stroke volume (SV), heart rate per minute (HR, min$^{-1}$), peripheral resistance (TPR), mean arterial pressure (MAP) and diastolic pulmonal-arterial pressure (PA$_d$).

With acetate dialysis, 3 patients experienced a symptomatic hypertensive episode. One of these patients also had a hypotensive episode during acetate hemodiafiltration in spite of a slight increase in the peripheral resistance (fig. 4).

The cardiac output increased in all bicarbonate procedures, usually due to a gain in stroke volume. With almost no exception the peripheral resistance decreased slightly under bicarbonate. This was also observed for the filtrative procedures. The mean arterial blood pressure at the end of the bicarbonate treatments was normal for all patients, no hypotensive episodes occurred (fig. 1−3).

Contrary to the acetate procedures, here the diastolic pulmonal-arterial pressure dropped adequately so that on one hand a preload relief occurred and on the other hand an afterload reduction due to constant or slightly decreasing peripheral resistance was observed. When comparing the cardiac outputs measured at the onset and the end of the procedures, and the diastolic pulmonal-arterial pressures as a pressure-flow relation in the Frank-Starling diagram to gain information on the changes in myocardial contractility, a downward shifting of the curves towards the lower right is found for acetate, whereas an upward shifting is found for bicarbonate, with the curves shifting primarily to the upper left (fig. 5).

The plasma bicarbonate concentrations as well as arterial $CO_2$ tension − except for those procedures affected by a symptomatic hypotension − were within the normal range. Usually, the normal range was reached faster with bicarbonate. For all acetate-buffered procedures the comparison of the catecholamine levels showed a more or less marked increase in epinephrine and norepinephrine levels, whereas with bicarbonate no changes or slightly decreased catecholamine levels were found (table I).


## Discussion

In HDA, both the stroke volume and the cardiac output were decreased at a usually insufficient increase in the heart rate. The peripheral resistance probably decreases due to the acidotic effect and the influence of the acetate upon the capillary bed. Consequently, the mean arterial pressure drops.

It is remarkable that the diastolic pulmonary-arterial pressure as an expression of the end-diastolic left-ventricular filling pressure [13, 20] did not normalize, and the heart function curves in all patients showed a de-

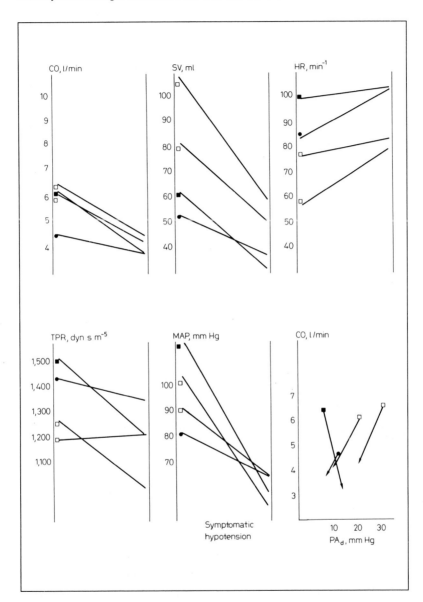

*Fig. 4.* Hemodynamic parameters during treatment with symptomatic hypotension. Patient I (□) during HDA and HDFA, patient IV (●) during HDA, and patient V (■) during HDA. Shown are cardiac output (CO), stroke volume (SV), peripheral resistance (TPR) and mean arterial pressure (MAP) during the course of treatment, and the shifting of the heart function curves ($PA_d$ dependent on CO).

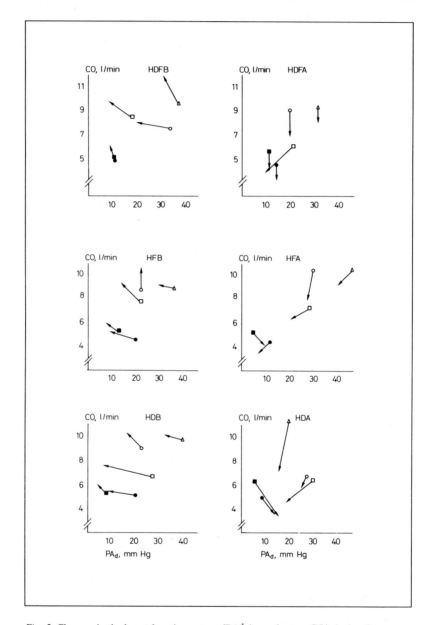

*Fig. 5.* Changes in the heart function curves (PA$^d$ dependent on CO) during the respective course of treatment, from the beginning in patient I (□), patient II (○), patient III (△), patient IV (●) and patient V (■), at the end of treatment in direction of the arrow. Left side shows HDFB, HFB, and HDB. Right side shows HDFA, HFA and HDA.

*Table I.* Serum levels of norepinephrine and epinephrine before and after 4 h of treatment with acetate procedures (HDA, HFA, HDFA) compared to bicarbonate procedures (HDB, HFB, HDFB)

|  | Norepinephrine, pg/ml | | Epinephrine, pg/ml | |
|---|---|---|---|---|
|  | before | after | before | after |
| HDA | 262 ± 63 | 472 ± 259 | 36 ± 15 | 66 ± 23 |
| HDB | 477 ± 249 | 375 ± 144 | 69 ± 30 | 56 ± 22 |
| HFA | 450 ± 210 | 634 ± 264 | 51 ± 17 | 62 ± 21 |
| HFB | 476 ± 229 | 416 ± 228 | 51 ± 12 | 48 ± 13 |
| HDFA | 310 ± 138 | 496 ± 225 | 48 ± 13 | 63 ± 12 |
| HDFB | 473 ± 171 | 364 ± 114 | 48 ± 21 | 56 ± 17 |

terioration in the pressure-volume relation, indicating a reduced contractility and compliance [14]. In extreme cases – here in 3 of 5 treatments – this clinically insufficient regulation of the circulation [14] led to a symptomatic hypotension which was also stated by other authors to range between 15 and 50% [18, 28, 30, 33]. A reason for the unfavourable cardiocirculatory regulation under HDA is the cardio-depressant effect of the acetate [1, 34, 36, 44, 60] which leads directly to vasodilation [1, 34, 35, 41, 55] and thus to a decrease in resistance. The metabolic acidosis [2, 22, 24] is not compensated due to the high loss of bicarbonate regeneration [39], and a high clearance of osmotically effective substances from the extracellular space which causes a volume shift towards the intracellular space and thus a reduction of the refilling, followed by an intravascular volume deficit [29]. The total of these negative effects upon the cardiocirculatory status exceed the classical vascular compensation mechanisms which normally provide protection against variations in the intravascular volume.

This is even more apparent in patients with chronic renal insufficiency where a damage of the afferent and possible also of the efferent reflex pathways by autonomic neuropathy may be present [11, 32, 37, 38, 51, 59], or an adequate catecholamine response does not occur because of qualitative and quantitative changes in the alpha-receptors [15, 65].

The vascular stability found in acetate filtrative procedures results from an increase in the peripheral resistance [5, 6, 9, 26, 33, 46, 58, 64]. Although stroke volume and cardiac output decrease as well, the normally adequate increase in the heart rate will limit the extent of the cardiac output reduction. Normally, the substantial increase in peripheral resistance will maintain the mean arterial pressure at a stable level. It is remarkable that the left-ventricular filling pressures usually remain increased, frequently even increasing from the normal range and thus causing an unfavorable change in the pressure-flow relation of the heart. The circulatory change observed here is mainly the consequence of a response by the sympathetic nervous system to compensate for a volume depletion with at least a partially adequate effect on the target organs [4, 16, 41, 55, 65].

The at times marked increase in the peripheral resistance, up to 50% of the initial value [58], can be compared to an afterload stress test [52, 53] and impairs the left-ventricular function. This could provide an explanation for the frequently increased diastolic pulmonal-aterial pressures and the unfavorable left-ventricular filling pressure conditions following HFA and HDFA.

An entirely different circulatory behaviour is observed with bicarbonate, despite a comparable weight loss during the treatment. Hemodynamic stability is achieved by increasing the cardiac output. Here it is important to note that with an unchanged or only slightly changed blood pressure behaviour the peripheral resistance usually undergoes a moderate decrease. The maintenance of the MAP under bicarbonate is ensured by an increase in the left-ventricular output and not by peripheral vasoconstriction. This mechanism could be demonstrated in all bicarbonate procedures and must thus be attributed to a direct bicarbonate effect. Circulatory effects which under acetate buffer with diffusive or convective modes gradually favour the filtrative procedures, are rendered unimportant by the effect the bicarbonate buffer. Here the cause is the direct positive inotropic effect of the bicarbonate. An improvement in the contractility and left-ventricular output has also been demonstrated in experiments [16, 17]. Invasive and echocardiographic studies confirm this hemodynamic behaviour with regard to the contractility and stroke volume in HDB [3, 7, 58]. The positive effect of bicarbonate upon the inotropic state of the myocardial cells is probably due to a rapid intracellular bicarbonate enrichment and consequently an H-ion decrease. Since this effect seems to be independent of the systemically measured H-ion concentration [1], an increased release of troponine formation sites for calcium ions is expected, independent of the extent of acidosis present at the onset of therapy, causing a better contractility by increasing the actin-myosin interaction [31, 45, 63]. With the additional presence of metabolic acidosis, the bicarbonate buffer will cause an increased release of calcium ions by the longitudinal system of the sarcoplasmatic reticulum [42, 45], showing a potentiating effect upon the contractility.

Despite a preload decrease by ultrafiltration, the above-mentioned bicarbonate effects upon the heart permit an increase in the cardiac output volume in all bicarbonate procedures. This mechanism of cardiac compensation requires less compensatory activity from the sympatho-adrenergic system, thus making malfunctions due to an autonomous neuropathy negligible. Membrane-specific effects upon the circulatory stability could not be demonstrated as shown in the comparison of our data with the results of hemodynamic studies using Cuprophan membranes [44, 49, 54, 60].

## References

1   Aizawa, Y.; Ohmori, T.; Imai, K.; Nara, Y.; Matsuoka, M.; Hirasawa, Y.: Depressant action of acetate upon the human cardio-vascular system. Clin. Nephrol. 8: 477 (1977).

2   Albertini, B. v.; Korpalani, A.; Goldstein, M.; Glabman, S.; Bosch, P.J.: Changes in $PCO_2$ during and after hemodialysis. Proc. Eur. Dial. Transplant. Forum 6: 199 (1976).

3   Aljama, P.; Martin-Malo, A.; Sanz, R.; Pasalodos, J.; Sancho, M.; Moreno, E.; Gomez, J.; Perez, R.; Burdiel, G.L.; Andres, E.: Left ventricular function during hemofiltration and hemodialysis. A comparative study. Proc. Eur. Dial. Transplant. Ass. 19: 281 (1982).

4   Baldamus, C.A.; Ernst, W.; Fassbinder, W.; Koch, K.M.: Differing hemodynamic stability due to differing sympathetic response. Comparasion of ultrafiltration, hemodialysis and hemofiltration. Proc. Eur. Dial. Transplant. Ass. 17: 205 (1980).

5   Baldamus, C.A.; Ernst, W.; Lysaght, M.J.; Shaldon, S.; Koch, K.M.: Hemodynamics in hemofiltration. Int. J. artif, Organs 6: 27 (1983).

6   Baldamus, C.A.; Knobloch, M.; Schoeppe, W.; Koch, K.M.: Hemodialysis/hemofiltration. A report of a controlled cross over study. Int. J. artif. Organs 3: 211 (1980).

7   Barcena, C.G.; Olivero, J.; Ayus, J.: Bicarbonate dialysis (BD) is hemodynamically superior to acetate dialysis (AD). Kidney int. 16: 881 (1980).

8   Bazzato, G.; Coli, U.; Landini, S.; et al.: Cold as cardiovascular stabilizing factor in hemodialysis. Hemodynamic evaluation (Abstract). Am. Soc. artif. internal Organs 12: 41 (1983).

9   Bergstroem, J.: Ultrafiltration without dialysis for removal of fluid and solutes in uremia. Clin. Nephrol. 9: 156 (1978).

10  Bergstroem, J.; Asaba, H.; Fuerst, P.; Oules, R.: Dialysis, ultrafiltration and blood pressure. Proc. Eur. Dial. Transplant. Ass. 13: 293 (1976).

11  Blömer, H.: Klinik der koronaren Herzerkrankung. Heutiger Stand. Medsche Welt 24: 1919 (1973).

12  Bosch, J.P.; MacMounte, F.; Albertinini, B. v.; Kahn, T.; Glabman, S.; Moutoussis, G.: Participation of red blood cells in bicarbonate transport across the dialyser. Trans. Am. Soc. artif. internal Organs 24: 343 (1980).

13   Bouchard, R.J.; Gault, J.H.; Roos, J., Jr.: Evaluation of pulmonary arterial enddias-
     tolic pressure as an estimate of left ventricular enddiastolic pressure in patients with nor-
     mal and abnormal left ventricular performance. Circulation *44:* 1072 (1971).

14   Braunwald, E.; Sonnenblick, E.H.; Ross, J., Jr.: Contraction of the normal heart; in
     Braunwald, Heart disease, a textbook of cardiovascular medicine, part II, chap. 12,
     p. 413 (Saunders, Phliadelphia 1980).

15   Brodde, O.E.; Daul, A.; Graben, N.: Decreased number of alpha-adrenergic receptors
     in platelets of patients on maintenance hemodialysis. Proc. Eur. Dial. Transplant. Ass.
     *19:* 270 (1982).

16   Canella, G.; Picotti, G.C.; Mione, G.; Christinelli, L.; Maiorca, R.: Blood pressure
     behaviour during dialysis and ultrafiltration. A pathogenic hypothesis on hemodialysis-
     induced hypontension. Int. J. artif. Organs *1:* 69 (1978).

17   Clancy, R.L.; Gingolani, E.E.; Taylor, R.R.; Graham, T.P., Jr.; Gilmore, J.P.: Influ-
     ence of sodium bicarbonate on myocardial performance. Am. J. Physiol. *212:* 917
     (1967).

18   Decoulet, P.; Proulx, J.; Aime, F.; Berger, C.; Bloch, P.; Goupy, F.; Legrain, M.:
     Programme dialyse-informatique, III. Données épidemiologiques, stratégies de dialyse
     et résultats biologiques. J. Urol. Nephrol. *82:* 101 (1976).

19   Dolan, M.J.; Wipp, B.J.; Dacidson, W.D.; Weitzmann, R.E.; Wassermann, K.:
     Hypopnea associated with acetate hemodialysis carbon dioxide flow-dependent ventila-
     tion. New Engl. J. Med. *305:* 72 (1981).

20   Falicov, R.E.; Resnekow, L.: Relationship of the pulmonary artery enddiastolic pres-
     sure to the left ventricular and diastolic and mean filling pressures in patients with and
     without left ventricular dysfunction. Circulation *42:* 65 (1970).

21   Guarnier, G.F.; Caretta, R.; Toigo, G.; Campanacci, L.: Acetate intolerance in chronic
     uremic patients. Nephron *24:* 212 (1972).

22   Hampl, H.: Einfluss des Säure-Basen-Status auf die Kreislaufstabilität während Azetat-
     und Bicarbonat-Hämodialysie; in Streicher, Schöppe, Die adäquate Dialyse. Dialyse-
     Ärzte Workshop Bernried 1981, p. 101 (Springer, Berlin 1982).

23   Hampl, H.; Klopp, H.W.; Gruber, W.M.; Pustelnik, A.; Schiller, R.; Kessel, M.;
     Hanefeld, F.: Zur Pathogenese der durch Azetatdialyse induzierten kardiovaskulären
     und zerebralen Probleme. Nieren-Hochdruck-Krankh. *12:* 132 (1983).

24   Hampl, H.; Paeprer, H.; Unger, V.; Ryzlewicz, T.; Fischer, C.; Cambi, V.; Kessel, M.:
     Hemodynamic studies, acid-base status and osmolality in different hemodialysis proce-
     dures. Artif. Organs *2:* 348 (1978).

25   Hampl, H.; Paeprer, H.; Unger, V.; Kessel, M.: Vergleichende hämodynamische Stu-
     dien bei konventioneller Hämodialyse mit vorangehender Ultrafiltration und Hämofil-
     tration. Nieren-Hochdruck-Krankh. *1:* (1978).

26   Hampl, H.; Paeprer, H.; Unger, V.; Fischer, C.; Resa, L.; Kessel, M.: Hemodynamic
     changes during hemodialysis, sequential ultrafiltration, and hemofiltration. Kidney int.
     *18:* 83 (1980).

27   Henderson, L.W.; Koch, K.M.; Dinoarello, C.A.; Shaldon, S.: Hemodialysis hypoten-
     sion: the interleukin hypothesis. Blood Purification *1:* 3 (1983).

28   Henderson, L.W.; Livotti, G.L.; Ford, C.A.; Kelly, A.; Lysaght, M.J.: Clinical experi-
     ence with intermittent hemodiafiltration. Trans. Am. Soc. artif. internal Organs *19:* 119
     (1973).

29   Hombrouckx, R.; Leroy, R.; Damme, W. van; Kliniek, M.D.: Four years bicarbonate

dialysis versus acetate dialysis: a medical long-term follow-up. (Abstract). Artif. Organs. *7:* 43 (1983).

30    Kant, K.S.; Pollack, V.E.; Chatey, M.; et al.: Multiple use dialyzers, safety and efficacy Kidney int. *19:* 729 (1981).

31    Kath, A.M.: Contractile proteins of the heart. Physiol. Rev. *50:* 63 (1970).

32    Kersh, E.S.; Kronfield, J.S.; Unter, A.; Popper, R.W.; Cantor, S.; Cohn, K.: Autonomic insufficiency in uremia as a cause of hemodialysis-induced hypotension. New Engl. J. Med. *290:* 650 (1974).

33    Keshaviah, P.; Ilstrup, K.; Constantini, E.; Berkseth, R.; Shapiro, F.: The influence of ultrafiltration (UF) and diffusion (D) on cardiovascular parameters. Trans. Am. Soc. artif. internal Organs *26:* 328 (1980).

34    Kirkendol, P.L.; Devia, C.J.; Holbert, R.D.: A comparison of the cardiovascular effect of sodium acetate, sodium bicarbonate and other potential sources of fixed base in hemodialysis solutions. Trans. Am. Soc. artif. internal Organs *23:* 399 (1977).

35    Kirkendol, P.L.; Robie, N.W.; Gonzales, F.M.; Devia, C.J.: Cardiac and vascular effects of infused sodium acetate in dogs. Trans. Am. Soc. artif. Organs *24:* 714 (1978).

36    Kirkendol, P.L.; Pearson, J.E.; Bower, J.D.; Holbert, R.D.: Myocardial depressant effects of sodium acetate. Cardiovasc. Res. *12:* 127 (1978).

37    Koch, K.M.; Ernst, W.; Baldamus, C.A.; Brecht, H.M.; Goerge, J.; Fassbinder, W.: Sympathetic activity and hemodynamics in hemodialysis, ultrafiltration and hemofiltration. Kidney int. *16:* 891 (1980).

38    Koch, K.M.; Baldamus, C.A.; Ernst, W.; Fassbinder, W.; George, J.; Brecht, H.M.: Autonome Kreislauffregulation in der Urämie. Klin. Wschr. *58:* 1037 (1980).

39    Kveim, M.; Nesbakken, R.: Utilization of exogenous acetate during hemodialysis. Trans. Am. Soc. artif. internal Organs *21:* 138 (1975).

40    Maggiore, Q.; Pizarelli, F.; Sisca, S.; et al.: Blood temperature and vascular stability during hemodialysis and hemofiltration. Trans. Am. Soc. artif. internal Organs *28:* 523 (1982).

41    Molnar, J.R.; Scott, J.B.; Fröhlich, E.D.; Haddy, F.J.: Local effects of various anions and $H^+$ on dog limb and coronary vascular resistances. Am. J. physiol. *203:* 125 (1962).

42    Nakamaru, Y.; Schwartz, A.: Possible control of intracellular calcium metabolism by ($H^+$). Sarcoplasmatic reticulum of skeletal and cardiac muscle. Biochem. biophys. Res. Commun. *41:* 830 (1970).

43    Nissenson, A.R.: Prevention of dialysis-induced hypoxemia by bicarbonate dialysis. Trans. Am. Soc. artif. internal Organs *26:* 339 (1980).

44    Novello, A.; Kelsch, R.; Easterling, R.: Acetate intolerance during hemodialysis. Clin. Nephrol. *5:* 28 (1976).

45    Poole-Wilson, P.A.; Langer, G.A.: Effect of pH on ionic exchange and function in rat and rabbit myocardium. Am. J. Physiol. *229:* 570 (1975).

46    Quellhorst, E.; Schuenemann, B.; Hildebrandt, U.; Falda, Z.: Response of the vascular system to different modifications of hemofiltration and hemodialysis. Proc. Eur. Dial. Transplant Ass. *17:* 197 (1980).

47    Quellhorst, E.B.; Schuenemann, B.; Hildebrandt, U.: How to prevent vascular instability: hemofiltration. Proc. Eur. Dial. Transplant Ass. *18:* 243 (1981).

48    Raja, R.M.; Henriquez, M.; Kramer, M.S.; Rosenbaum, J.L.: Improved dialysis tolerance using Redy sorbent system with bicarbonate dialysate in critically ill patients. Dial. Transplant *8:* (1979).

49 Raya, R.M.; Kramer, M.S.; Rosenbaum, J.L.; Bolisay, C.; Krug, M.: Prevention of hypotension during iso-osmolar hemodialysis with bicarbonate dialysate. Trans. Am. Soc. artif, internal Organs *26:* 375 (1980).

50 Ritz, E.; Bosch, J.; Henderson, L.W.; Kishimoto, T.; Koch, K.M.; Pierdides, A.; Shaldon, S.; Streicher, E.: Hemofiltration and vascular stability. Contr. Nephrol. *32:* 200 (1982).

51 Romoff, M.S.M.; Mcampese, V.; Lane, K.; Massary, S.G.: Mechanism of autonomic dysfunction in uremia: evidence for reduced and organ response to norepinephrine. Kidney int. *14:* 731 (1978).

52 Roos, J., Jr.; Braunwald, E.: The study of left ventricular function in man by increasing resistance to ventricular ejection with angiotensin. Circulation *29:* 739 (1964).

53 Ross, J.,Jr.; Covell, J.W.; Sonnenbuck, E.H.; Braunwald, E.: Contractile state of heart characterized by forcevelocity relations in variability afterload and isovolumic beats. Circulation Res. *18:* 149 (1966).

54 Rouby, J.J.; Rottembourg; Durande, J.P.; Basset, J.Y.; Degoulet, P.; Glaser, P.; Legrain, M.: Hemodynamic changes induced by regular hemodialysis and sequential ultrafiltration hemodialysis, a comparative study. Kidney int. *17:* 801 (1980).

55 Santoro, A.; Charini, C.; Esposti Degli, E.; Sturawi, A.; Zuccala, A.; Zucjelli, P.: Effects of hemofiltration (HF) and hemodialysis (HD) on autonomic control of circulation. Blood Purification *1:* 55 (1983).

56 Sargent, J.A.; Gotsch, A.F.: Bicarbonate and carbon dioxide transport during hemodialysis. Am. Soc. artif. internal Organs J. *2:* 61 (1979).

57 Schneider, H.; Streicher, E.: Transport characterisation of a new polysulfone membrane. Artif. Organs (in press).

58 Shaldon, S.; Beau, M.C.; Deschodt, G.; Ramperez, P.; Mion, C.: Vascular stability during hemofiltration. Trans. Am. Soc. artif. internal Organs *26:* 391 (1980).

59 Tomiyama, O.; Shiigai, T.; Tomita, K.; Mito, Y.; Shinohara, S.; Takeuchi, J.: Baroreflex sensivity in renal failure. Clin. Sci. *58:* 21 (1980).

60 Vincent, J.L.; Herwegheim, J.L. van; Degaute, J.-P.; Berre, J.; Dufaye, P.; Kahn, R.J.: Acetate-induced myocardial depression during hemodialysis for acute renal failure. Kidney int. *22:* 653 (1982).

61 Wang, H.H.; Katz, R.L.: Effects of changes in coronary blood pH on the heart. Circulation Res. *17:* 114 (1965).

62 Wehle, B.; Asaba, H.; Castenfors, J.; Fuerst, P.; Gunnarson, B.; Shaldon, S.; Bergstroem, J.: Hemodynamic changes during sequential ultrafiltration and dialysis. Kidney int. *18:* 411 (1979).

63 Williamson, J.R.; Safer, B.; Rich, T.; Schaffer, S.; Kabayashi, K.: Effects of acidosis on myocardial contractility and metabolism. Acta med. scand. *587:* 95 (1976).

64 Wizemann, V.; Rawer, P.; Schütterle, G.: Ultrashort hemodiafiltration: long-term efficiency and hemodynamic tolerance. Proc. Eur. Dial. Transplant Ass. *19:* 175 (1982).

65 Wizemann, V.; Lubbecke, F.; Konrad, C.; Schütterle, G.: Sympathetic and organ regulation during different methods of renal replacement (Abstract). Blood Purification *1:* 60 (1983).

Dr. H. Schneider, Abteilung für Nieren- und Hochdruckkrankheiten,
Zentrum für Innere Medizin, Katharinenhospital, Kriegsbergstrasse 60,
D-7000 Stuttgart 1 (FRG)

Contr. Nephrol., vol. 46, pp. 151–161 (Karger, Basel 1985)

# Hemodynamic Studies in Chronic Dialysis Patients with a Polysulfone Hemodiafilter

*H. Jahn, M. Bauler, D. Schohn*

Service de Néphrologie, ULP, Strasbourg, France

## Introduction

The extracorporeal blood circulation used in the maintenance treatment of end-stage renal failure my be a cause of adverse reactions in the patients, among which hemodynamic instability has a prominent place.

The common factors responsible for producing cardiocirculatory disturbances are: (1) the modifications of the blood volume induced by ultrafiltration; (2) the composition of the dialysate, especially acetate or bicarbonate; (3) the biocompatibility of the extracorporeal circuit specifically the membrane of the dialyzer.

Any part of the cardiovascular system may be involved: myocardium and vessels, low pressure system and high pressure system.

To analyze the effects on hemodynamics of the new highly permeable polysulfone membrane (Fresenius F 60) we applied the standardized hemodynamic investigations used in our unit [11].

## Methodology

The standardized investigations comprise the study of the low pressure system, the high pressure system, and the cardiac function, and are described elsewhere [6, 11].

The good tolerance of the F 60 hemodiafilter allowed us to apply this filter in routine treatments. The examinations were performed in patients who could be adapted to new treatment modalities in order to allow us to perform hemodynamic investigations: (1) 10 chronic hemodialysis patients, mean age 48 ± 11 years were studied during hemodialysis, hemofiltration and hemodiafiltration, the F 60 being used in all cases. (2) 6 chronic hemodialysis patients, mean age 44 ± 7 years, were studied during bicarbonate hemodialysis without fluid removal, again with F 60. (3) 5 chronic hemodialysis patients, mean age 51 ± 7 years, were studied during the use of the F 60 in an acetate versus, bicarbonate hemodialysis; no fluid was removed during the first 3 h of hemodialysis in order to study the effects of the membrane per se on the hemodynamic parameters. Fluid was thereafter removed during the next 3 h of treatment. In comparison to a Cuprophan dialyzer and to a polyamid dialyzer the modifications of $PaO_2$, $PaCO_2$ and biocompatibility were examined for their effects in these patients.

In one case of hypertension the F 60 hemodialysis was compared to a standard Cuprophan hemodialysis.

Informed consent was obtained from each patient.

### Treatment Modalities

We used as dialyzer the Hemoflow C 1.0 with a Cuprophan membrane (Fresenius, FRG) and as hemodiafilter the Hemoflow F 60 with a polysulfone membrane (Fresenius, FRG). In hemodialysis we used acetate-containing dialysate (acetate 38 mEq/l, Na 140 mEq/l, Ca 3.5 mEq/l, and bicarbonate-containing dialysate (bicarbonate 37 mEq/l, Na 140 mEq/l, Ca 3.5 mEq/l). In hemofiltration we exchanged 20 liters per treatment, monitor: HFM® (Gambro, Sweden), the substitution fluid contained acetate 37 mEq/l, Na 140 mEq/l, Ca 3.5 mEq/l. For hemodiafiltration we used substitution fluid with the composition mentioned above. Hemodialysis and hemodiafiltration were performed with A 2008 HDF monitor (Fresenius, FRG).

### Biochemistry

The usual biological parameters and plasma renin activity were measured before and after each treatment.

During the treatment, $PaO_2$ and $PaCO_2$ were monitored by a continuous blood gas analyzer through an intravascular catheter (Odam-Brucker Lab.).

The biocompatibility was assessed by the modification of the blood white cell count, by the plasma level of total complement ($CH_{50}$) and its fractions C3 and C4 and by the thrombocyte count.

The results are expressed as mean ± standard deviation. The statistical signification is assessed by the paired Student's t test.

### Results

### The Low Pressure System

The pressure variations in the low pressure system are dependent on the blood volume modifications [4, 6, 8, 11, 14]. The analysis of specific effects of the F 60 on the low pressure system, performed during bicarbonate dialysis without fluid removal, shows that the pulmonary wedge pressure (PWP) remains stable during the whole hemodialysis treatment procedure, as the right atrial pressure. This means that the filling pressure of left and right ventricles are not affected by the F 60 per se (fig. 1).

As bicarbonate alone has no significant effects on the hemodynamic parameters, it appeared necessary to explore the effects on the PWP stability of both acetate and bicarbonate hemodialysis without and then with fluid removal. When no fluid is removed PWP remains stable both with

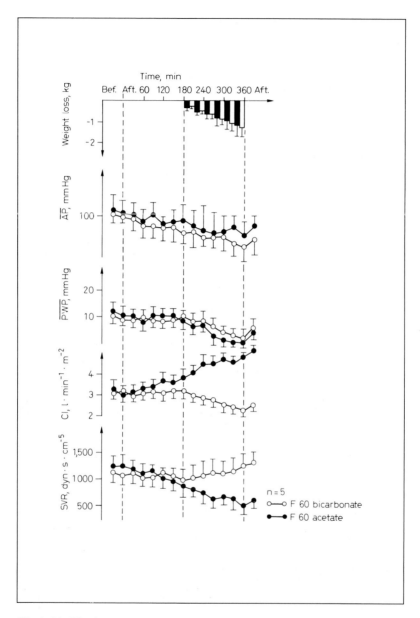

*Fig. 1.* Modification of the arterial pressure (AP), the pulmonary wedge pressure (PWP), the cardiac index (CI) and the systemic vascular resistances in 5 chronic hemodialysis patients (CHP) during an F 60 acetate versus bicarbonate hemodialysis. Each treatment comprises two periods: one without fluid removal and one with fluid removal.

acetate and bicarbonate, whereas during fluid removal PWP decreases. This decrease is, however, more important with acetate.

As the F 60 is a highly permeable membrane with therefore high ultra-filtration rates, the limits of ultrafiltration per time unit in regard to PWP pressure decrement must be specified. We have demonstrated that below a weight loss of 3% of body weight only small PWP modifications occur [8]. A weight loss of 6% of body weight induces a PWP decrement of near 10 mm Hg. Therefore, with a weight loss greater than 6% of body weight important PWP decrements can be observed and can have deleterious effects on the left ventricular filling pressure and so on the left ventricular function. This occurs particularly when the initial PWP value is within the normal range. Nevertheless, the end result on blood pressure depends also on the autonomic regulation of heart and vessels.

It is now of interest to compare the effects of hemodialysis, hemodiafiltration and hemofiltration with the same F 60 on PWP. For a comparable weight loss the PWP decrease is less important during hemofiltration and hemodiafiltration than during hemodialysis. Therefore, with a same hemodiafilter (F 60) one can observe a difference in the treatment modalities (fig. 2).

As the role of the left ventricular filling pressure is one of the determinants of the hemodynamic stability, we would like to give an overview of the factors influencing the relation between fluid removal and PWP decrement, thus giving indications for the management of ultrafiltration: (1) Vascular refilling may depend on capillary permeability and can be increased by blood osmolality. Increased Na levels in the dialysate enhance the vascular refilling [1, 9, 10]. (2) The compliance of the low pressure system is also an important component. However, the mechanisms acting on the compliance are less definable. But ionic composition of blood ($Na^+$, $Ca^{++}$), hormone levels and vascular receptor efficiency may be implied [7]. (3) The ventricular performance when reduced as in myocardiopathy or during β-blocker treatment may induce great pressure (PWP) decrements with small weight loss [8].

### The High Pressure System

The high pressure system comprises the left ventricle and the resistance vessels [7, 11]. This system is mainly influenced by first, acetate dialysate, second, the hormones acting on heart vessels (we must mention that the highly permeable filter can interfere with plasma hormone levels), and, third, the integrated reactions to volume depletion [6, 11].

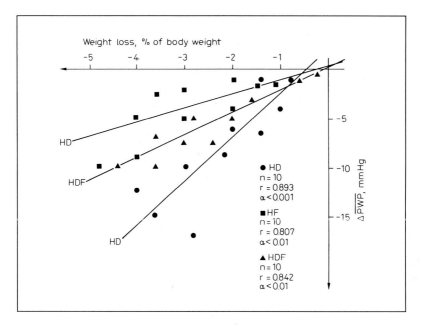

*Fig. 2.* Relation between the pulmonary wedge pressure decrement (△ PWP) and the weight loss (expressed in % of initial body weight) in 10 chronic hemodialysis patients during hemofiltration (HF), hemodiafiltration (HDF) and hemodialysis (HD).

With the F 60, we observed under acetate and bicarbonate dialysis a similar behaviour of the systemic vascular resistances as we observed with a standard Cuprophan dialyzer, already described [7, 11–13].

Acetate dialysate induces a decrease of systemic vascular resistances, whereas with bicarbonate we observe no modifications of the systemic vascular resistances in the absence of fluid removal and an increase of the systemic vascular resistances when fluid is removed by ultrafiltration, according to the integrated reactions to volume depletion [6, 8, 13].

A symptomatic hypotension takes place with acetate when cardiac output does not compensate the decrease of the systemic vascular resistances. With both bicarbonate and acetate dialysis performed with the F 60, we did not observe symptomatic hypotension in the studied patient groups, but we must mention that these patients had neither signs of myocardiopathy nor of autonomic dysfunction.

We would like to add one observation in a patient with hypertension showing that there may be a possible effect of the membrane through its

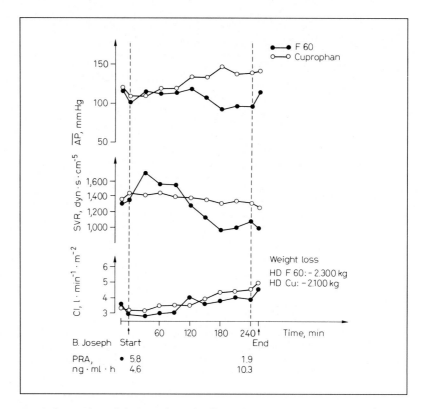

*Fig. 3.* Comparison of the hemodynamic effects of an F 60 acetate hemodialysis versus an acetate hemodialysis with a standard Cuprophan dialyzer in a hypertensive chronic hemodialysis patient. PRA = Plasma renin activity.

sieving coefficient (fig. 3). This patient displayed hypertension and the blood pressure continued to increase during each hemodialysis procedure with a standard Cuprophan dialyzer; whereas in the F 60 hemodialysis blood pressure decreased. With Cuprophan dialysis, plasma renin activity increased; with the F 60 dialysis plasma renin activity decreased [9].

### Cardiac Performance

When we evaluate cardiac performance with Franck Starling curves we observe that cardiac output increases during acetate hemodialysis with or without fluid removal, whereas cardiac output decreases during bicarbonate hemodialysis, only when fluid is removed [6, 7, 12].

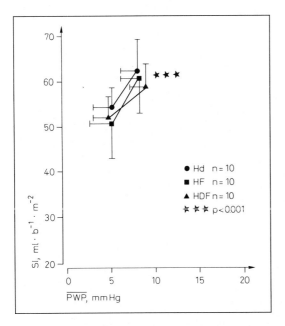

*Fig. 4.* Cardiac function curves obtained in 10 chronic hemodialysis patients after hemodialysis (HD), hemofiltration (HF) and hemodiafiltration (HDF) with the F 60 polysulfone membrane.

The study performed with the F 60 showed similar results. With this technique we could not observe any effect of the membrane on the function curves, nor did we observe different behaviors of the function curves when comparing the effects of hemodialysis, hemofiltration and hemodiafiltration (fig. 4).

### Pathogenetic Considerations

*Acetate.* One would have expected that high permeability filters could influence the transfer of acetate toward the patient. The determination of plasma acetate levels have, nevertheless, not shown different plasma values that those observed with other membranes used in dialyzers of a similar size [13].

*Blood Gases.* A permanent analysis of the blood gases has shown that we have an initial decrease of $PaO_2$ with the Cuprophan membrane, a de-

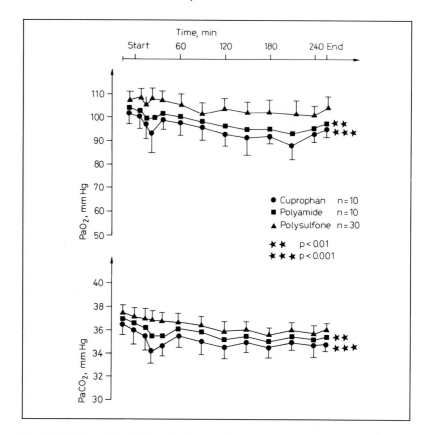

*Fig. 5.* Evolution of the PaO$_2$ and of the PaCO$_2$ in chronic hemodialysis patients during hemodialysis with a polysulfone membrane (F 60), a polyamide membrane and a Cuprophan membrane.

crease which does not occur with the F 60 (fig. 5). Afterwards, the overall modifications of PaO$_2$ and PaCO$_2$ with acetate are comparable to the data of the literature [2, 3, 5, 11].

*Indicators of Biocompatibility.* With the F 60, we did not observe the usually described leukocyte drop, change in thrombocytes and complement activation that occur with Cuprophan [2, 3] (fig. 6). Therefore, one can speculate that the absence of complement activation avoids the pulmonary leukostasis and its consequences on blood gases and perhaps on pulmonary circulation.

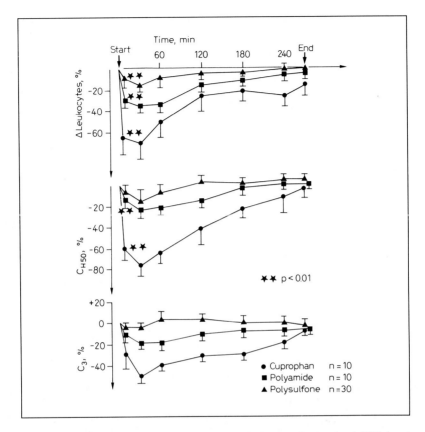

*Fig. 6.* Evolution of the leukocytes, of the total plasma complement level (CH$_{50}$) and of the C3 fraction in chronic hemodialysis patients during hemodialysis with a polysulfone membrane (F 60), a polyamide membrane and a Cuprophan membrane.

## Conclusions

We applied standardized hemodynamic investigations to study the hemodynamic effects of the F 60 polysulfone hemodiafilter comprising of studies of the low pressure system, the high pressure system, the cardiac function and the blood gases modifications.

This filter may act on hemodynamics by its high permeability properties and by the nature of the polysulfone membrane.

The low pressure system is mainly influenced by the fluid removal. However, when compared to Cuprophan, the F 60 polysulfone membrane by its better biocompatibility may have fewer effects on the pulmonary circulation.

The resistance vessels are mainly affected by the type of buffer salt used. The acetate load for the patients seems to be the same with the F 60 as in a standard Cuprophan dialysis. Nevertheless, the high sieving coefficient of the membrane may act on plasma vasoactive hormone levels and therefore influence blood pressure.

Cardiac output is increased by acetate and seems not to be influenced by the polysulfone membrane.

The cardiac function curves are influenced both by the fluid removal and by acetate. The F 60 polysulfone membrane seems not to affect the cardiac function curves per se.

The high permeability of this new membrane allows its use in hemofiltration, hemodialysis and hemodiafiltration with the A 2008 machine which permits the most convenient treatment for each patient.

## References

1   Brunois, J.P.; Toupance, O.; Vistelle, R.; Choisy, H.; Chanard, J.: Sodium dependent transcellular water shift during dialysis. I. Change in extracellular fluid volume. Nephrology 5: 27 (1984).

2   Craddock, P.G.; Fehr, J.; Dalmasso, A.P.; Brigham, K.L.; Jacobs, H.J.: Pulmonary vascular leucostasis resulting from complement activation by dialyser cellophane membranes. J. clin. Invest. 59: 879 (1977).

3   De Baker, W.A.; Verpooten, G.A.; Borgonjon, D.J.; Van Waelegheim, J.P.; Vermeire, P.A.; De Broe, M.E.: Hypoxemia during hemodialysis. Effect of different membrane and dialysate composition. Contr. Nephrol., vol. 37, p. 134 (Karger, Basel 1984).

4   Gauer, O.H.: Role of the cardiac mechanoreceptors in the control of plasma volume; in Osmotic and volume regulation. Alfred Benson symposium XI, p. 224 (Munksgaasd, Copenhagen 1978).

5   Hampl, H.; Klopp, H.W.; Pustelnick, A.; Wolff, P.; Gruber, M.; Schiller, R.; Kessel, M.: Zur Pathogenese der durch Azetatdialyse induzierten kardiovaskulären und zerebralen Probleme; in Niereninsuffizienz. Seybold, Schulz, Pilgrim, Aktuelle klinische und therapeutische Probleme, p. 107 (Dustri, München 1981).

6   Jahn, H.; Schohn, D.; Schmitt, R.: Etudes hémodynamiques au cours de l'insuffisance rénale terminale. Effets des techniques d'épuration extra-rénale. Nephrology 2: 53 (1981).

7   Jahn, H.; Schohn, D.; Schmitt, R.; Moch, S.: Azetat und Bikarbonat Dialyse. Hemodynamische Wirkungen; in Niereninsuffizienz. Seybold, Schultz, Pilgrim, Aktuelle klinische und therapeutische Probleme, p. 87 (Dustri, München 1983).

8   Jahn, H.; Schohn, D.; Schmitt, R.: Hemodynamic modifications induced by fluid removal and treatment modalities in chronic hemodialysis patients. Blood Purification 1: 80 (1983).

9     Jahn, H.; Schohn, D.; Gullberg, C.; Schmitt, R.: Hemodynamic long-term effects of hemofiltration on dialysis and drug-resistant hypertension. Contr. Nephrol., vol. 32, p. 61 (Karger, Basel 1982).

10   Rouby, J.J.; Rottembourg, J.; Durrande, J.P.; Basset, J.Y.; Legrain, M.: Plasma volume changes induced by regular hemodialysis and controlled sequential ultrafiltration hemodialysis. Dial. Transplant. *8:* 237 (1979).

11   Schohn, D.; Gessler, W.; Schmitt, R.; Jahn, H.: Einfluss der dialysatzusammensetzung auf die Hämodynamik Nieren-Hochdruck-Krankh. (to be published, 1984).

12   Schohn, D.; Jahn, H.; Doerflinger, C.; Hashimoto, T.: Increase of stroke index by acetate hemodialysis. Contr. Nephrol., vol. 41, p. 383 (Karger, Basel 1984).

13   Schohn, D.; Klein, S.; Mitshuishi, Y,; Jahn, H.: Correlation between plasma sodium acetate level and systemic vascular resistances. Proc. Eur. Dial. Transplant Ass. *18:* 160 (1981).

14   Wehle, B.; Asaba, H.; Castenfors, J.; Furst, P.; Grahn, A.; Gunnarson, B.; Shaldon, S.; Bergstrom, J.: The influence of dialysis fluid composition on the blood pressure. Clin. Nephrol. *10:* 62 (1978).

Prof. H. Jahn, Service de Néphrologie, Hospices Civils de Strasbourg,
1, Place de l'Hôpital, F-67005 Strasbourg Cédex (France)

# Clinical Applications

Contr. Nephrol., vol. 46, pp. 162–168 (Karger, Basel 1985)

# Hemodiafiltration – a Superior Method of Blood Purification in Children?

*M. Fischbach, C. Koehl, J. Geisert*

Service de Pédiatrie, Hôpital de Hautepierre, Strasbourg, France

## Introduction

Children with end-stage renal failure should be treated with a blood purification method which, whilst providing the optimal purification, requieres the least possible time per session and nevertheless avoids the intolerance symptoms associated with a very rapid weight loss.

A blood purification procedure used in children should on the one hand ensure the elimination of small molecules (urea, creatinine, phosphate) – primarily by diffusion during a hemodialysis (HD) session – and, on the other hand, ensure also the elimination of the middle molecules – primarily by ultrafiltration in a hemofiltration (HF) session. The new therapeutical mode of hemodiafiltration (HDF) [1] consists of a combination procedure for the simultaneous elimination of small and middle molecules during one session.

## Technique of Hemodiafiltration in Children

For approximatively 2 years now [2], the HDF method has been used to take over the kidney function in our children with chronic end-stage renal failure. We have used Fresenius A 2008 ABG machines which are also used for adult patients without any modifications. This machine is equipped with an automated balance for the automatic control of the substitution of removed filtrate.

*Table I.* Hemodiafilters for children

| | Fresenius | | Asahi | | Hospal | |
|---|---|---|---|---|---|---|
| | F 40 | F 60 | CS ultrafilter | PAN 150 | 1200 S | 1800 S |
| Membrane material | polysulfone | polysulfone | PAN | PAN | PAN | PAN |
| Area, m$^2$ | 0.65 | 1.25 | 0.5 | 1.0 | 0.5 | 0.7 |
| Blood volume, ml | 45 | 75 | 35 | 70 | 52 | 72 |
| Residual blood volume, ml | < 1 | < 1 | > 1 | > 1 | > 1 | > 1 |

The membrane has to be adapted for use in pediatric patients. The selection of the membrane is a very important criterion for the solute removal. Apart from the requirements for low blood priming volume, low residual blood volume and biocompatibility, parallel flow (plate) dialyzers (Hospal: Biospal 1200 S, Biospal 1800 S) as well as capillary dialyzers (Fresenius Hemoflow F 40, Asahi Ultrafilter CS, PAN 150) should also avail an increased UF coefficient which permits an adequate filtrate flow. The clearance values for small and middle molecules should ensure that the speed at which the removal of the uremic toxins occurs will be well tolerated by the child's organism (osmotic syndrome) (table I).

Blood flows reached in children range from 90 to 150 ml/min. These low blood flows are the limiting factor in the use of HF in children, since in order to achieve a sufficient elimination of urea, HF requires a longer duration of the session or higher blood flows. However, under HDF conditions, an effective filtration (regarding solute elimination, clinical tolerance and circulatory stability) can be achieved at the low blood flow usually encountered in children. The level of the filtrate flow is calculated from the blood flow present. The formula 'filtrate flow = 1/3 of blood flow' (table II) can in practice be applied to every mode of treatment, independent of the shunt conditions. Studies of the maximum filtrate flow which can be achieved in children [3] showed that a 25% increase in the filtrate flow resulted in an impressive 360% increase in the transmembrane pressure. The effects of an increased TMP such as marked thickening of blood, hemolysis, coagulation, $CO_2$ diffusion and high residual blood volume must be carefully weighed against the clinical requirements such as an increased elimination of middle molecules. During a HDF session of 3 h (table III), a filtrate volume is reached (filtrate flow = 1/3 of blood flow) that exceeds the total blood volume by four times. In comparison to adults, the relation between filtrate volume/blood volume in HDF is markedly increased in children.

Especially in children, the ultrafiltration should be exactly adjustable in order to avoid a drop in blood pressure. In HDF, the exact setting of the weight loss is ensured. This is one of the reasons for the excellent clinical tolerance of this procedure. The possible weight loss amounts to 2.5% of the body weight per hour, which means in children with a dry body weight from 15 to 17 kg a weight loss of 0.3 kg/h, with 17–22 kg a loss of 0.4 kg/h and with 22–25 kg a loss of 0.5 kg/h, respectively. HDF as a high-flux method also indicates certain risks – cardiovascular problems (importance of the volume exchanged), osmotic syndrome secondary to a very rapid elimination of solutes (especially small molecules) or depleting the organism of substances of physiological importance (trace elements, vitamins, hormones).

*Table II.* Blood fluid versus filtration in HDF with the polysulfone membrane (F 40).

| | $\dot{Q}_B$, ml/min | | |
|---|---|---|---|
| | 90 | 120 | 150 |
| $\dot{Q}_{UF} = 1/3\ \dot{Q}_B$, ml/min<br>PTM-150 ± 30, mm Hg | 30 | 40 | 50 |
| $\dot{Q}_{UF}$ max, ml/min<br>PTM-550, mm Hg | 38 ± 5 | 47 ± 3 | 56 ± 5 |

*Table III.* Relation between UF volume and circulatory compartment

| $\dot{Q}_B$,<br>ml/min | $\dot{Q}_{UF}$,<br>ml/min | UF volume per session,<br>ml/3 h | Body weight,<br>kg | Circulatory compartment<br>80 ml/kg |
|---|---|---|---|---|
| 90 | 30 | 5,400 | 15−17 | 1,200−1,360 |
| 120 | 40 | 7,200 | 17−22 | 1,360−1,760 |
| 150 | 50 | 9,000 | 22−25 | 1,760−2,000 |

## *Studies and Results* (Review)

We performed a comparative study in a group of children (average age 9 years, average weight 21 kg). In the first year, conventional hemodialysis (HD) was performed (plate dialyzer 'Höltzenbein', Travenol), followed by HDF in the next year (capillary dialyzer PAN 150, Asahi). Our studies confirmed the results obtained in adults. With HDF the duration of the treatment session was reduced by 30−40%, whereas the values of urea and creatinine elimination remained identical to those achieved by HD (table IV). HDF was well tolerated by the children. Phosphate binders were not required (reduced risk of aluminium intoxication) and we also found that less transfusions were needed [2].

The solute elimination possible with HDF depends to a great extent on the selected membrane [3].

*Table IV.* Relationship between serum levels of urea, creatinine, phosphate in HD and HDF (at start and end of treatment session) in 6 children

|  | HD (15 h/week) | | HDF (7.5−9 h/week) | |
|---|---|---|---|---|
|  | start | end | start | end |
| Urea, mmol/l | 40 ± 6 | 16 ± 3 | 32 ± 4 | 6 ± 2 |
| Creatinine, μmol/l | 1,080 ± 256 | 410 ± 167 | 810 ± 280 | 240 ± 123 |
| Phosphate, mmol/l | 1.65 ± 0.28 | 0.9 ± 0.19 | 1.34 ± 0.15 | 0.5 ± 0.08 |

5 children, average age 9 years (7−11), average weight 20.75 kg (20.5−21.0), average Hct of 24% (22−26%) were treated with a high−flux HDF (Fresenius Hemoflow F 40) during HD, HF and HDF. For comparison, they also underwent HF using 4 different membranes (table I). The blood flow was 120 ml/min, dialysate flow 500 ml/min (for HD/HDF), filtrate flow 40 ml/min (HDF/HF). Determinations of the extraction coefficient (E), sieving coefficient (S) and clearances were made 30 and 150 minutes after the onset of the treatment. Arterial and venous hematocrits were measured. Urea (MW 60), creatinine (MW 113), phosphate, $\beta_2$-microglobulin (MW 11,900), lysozyme (MW 1,6000), 25-OH-vitamin D3 (MW 25,000) were determined in the arterial part, the venous part (taking into account the thickening of the blood, corrections based on hematocrit) and in the filtrate. 5 ml of Inutest were administered 15 and 135 min after the onset of the treatment, and the presence of Inutest in arterial and venous blood as well as in the filtrate was measured after 30 and 150 min.

The results are summarised in tables V, VI and VII. In HDF the highly permeable membrane of the Hemoflow F 40 shows high extraction coefficients for small molecules at low blood flows. This is associated with the risk of a too rapid elimination of urea (osmotic syndrome) which should be met by slowly establishing an adaptation phase over several treatment sessions. The clearance reached in HDF does not equal the sum of the single clearances reached with HD and HF (table VI), since diffusion and convection take place simultaneously and mutually affect each other.

Compared to HD, the gain in elimination achieved by HDF increases especially by filtration, particularly for substances of various molecular weights. HDF with a highly permeable membrane ensures an optimal elimination of all uremic toxins (small, middle and large molecules). The selection of the proper membrane is decisive, especially for the elimination of higher molecular substances (table VII).

Depending on the membrane selected, the sieving coefficient drops to zero for substances of different molecular weights. For Asahi's polyacrylonitrile membrane (PAN 150, Ultrafilter CS) this substance is $\beta_2$-microglobulin, for Hospal's polyacrylonitrile (Biospal 1200 S, Biospal 1800 S) it is lysozyme, for Fresenius' polysulfone (Hemoflow F 40) it is 25-OH-vitamin D3. This heterogenity of various membranes is not important for small molecules such as urea, but it is probably of great importance of the treatment and the clinical effectivity of the solute elimination.

*Table V.* Extraction coefficient (E) (mean ± 2 SD) versus molecular weight compounds for the polysulfone membrane (F 40) at 30 and 150 min session times, in HD, HF and HDF

| | Time, min | Substance | | | | | | |
| --- | --- | --- | --- | --- | --- | --- | --- | --- |
| | | urea | creatine | phosphate | poly-fructosan | $\beta_2$-micro-globulin | lysozyme | 25-OH-vitamin $D_3$ |
| $E_{HD}$ | 30 | 0.90 ± 0.03 | 0.75 ± 0.05 | 0.74 ± 0.02 | 0.17 ± 0.01 | 0.13 ± 0.01 | 0.068 ± 0.004 | 0.023 ± 0.003 |
| | 150 | 0.85 ± 0.04 | 0.68 ± 0.03 | 0.67 ± 0.01 | 0.15 ± 0.01 | 0.12 ± 0.01 | 0.059 ± 0.003 | 0.014 ± 0.002 |
| $E_{HF}$ | 30 | 0.33 ± 0.01 | 0.34 ± 0.02 | 0.32 ± 0.04 | 0.32 ± 0.02 | 0.19 ± 0.01 | 0.11 ± 0.03 | 0.05 ± 0.009 |
| | 150 | 0.32 ± 0.01 | 0.31 ± 0.01 | 0.28 ± 0.03 | 0.28 ± 0.03 | 0.17 ± 0.02 | 0.10 ± 0.01 | 0.03 ± 0.004 |
| $E_{HDF}$ | 30 | 0.930 ± 0.01 | 0.85 ± 0.05 | 0.87 ± 0.02 | 0.55 ± 0.015 | 0.31 ± 0.02 | 0.18 ± 0.02 | 0.07 ± 0.002 |
| | 150 | 0.92 ± 0.02 | 0.79 ± 0.01 | 0.83 ± 0.03 | 0.53 ± 0.02 | 0.27 ± 0.01 | 0.15 ± 0.01 | 0.04 ± 0.006 |

*Table VI.* clearances (K: ml/min) versus molecular weight compounds (daltons) for the polysulfone membrane (F 40) at 30 and 150 min session times, in HD, HF and HDF

| | Time, min | Substance | | | | | | |
| --- | --- | --- | --- | --- | --- | --- | --- | --- |
| | | urea | creatinine | phosphate | poly-fructosan | $\beta_2$-micro-globulin | lysozyme | 25-OH-vitamin $D_3$ |
| $K_{HD}$ | 30 | 108 ± 3.6 | 90 ± 6 | 89 ± 2.4 | 20.4 ± 1.2 | 15.6 ± 1.2 | 8.2 ± 0.48 | 2.27 ± 0.4 |
| | 150 | 102 ± 4.8 | 82 ± 3.6 | 80 ± 1.2 | 18 ± 1.2 | 14.4 ± 1.2 | 7.0 ± 0.36 | 1.09 ± 0.1 |
| $K_{HF}$ | 30 | 39.6 ± 1.2 | 40.8 ± 2.4 | 38.4 ± 4.8 | 38.4 ± 2.4 | 22.8 ± 1.2 | 13.2 ± 3.6 | 8.02 ± 0.3 |
| | 150 | 3.4 ± 1.2 | 37.2 ± 1.2 | 33.6 ± 3.6 | 33.6 ± 3.6 | 20.4 ± 2.4 | 12 ± 1.2 | 6.93 ± 0.2 |
| $K_{HDF}$ realized | 30 | 111 ± 1.2 | 102 ± 6 | 104 ± 2.4 | 66 ± 1.8 | 37.2 ± 2.4 | 21.6 ± 2.4 | 10.04 ± 0.4 |
| | 150 | 110 ± 2.4 | 94.8 ± 1.2 | 99.6 ± 3.6 | 63.6 ± 2.4 | 32.4 ± 1.2 | 18 ± 1.2 | 10.07 ± 0.1 |
| $K_{HDF}$ calculated | 30 | 112 | 100.2 | 98.9 | 52.2 | 35.4 | 20.5 | 9.97 |
| | 150 | 107 | 93.8 | 91.2 | 46.6 | 32.3 | 18.3 | 9.92 |

Table VII. Sieving coefficient (S) (mean ± SD) versus molecular weight compounds (daltons), of different at 30 and 150 min session times, in HF

| | Time, min | Substance | | | | | | |
|---|---|---|---|---|---|---|---|---|
| | | urea | creatinine | phosphate | polyfructosan | $\beta_2$-microglobulin | lysozyme | 25-OH-vitamin $D_3$ |
| $F_{40}$ | 30 | 1.002 | 1.08 | 1.04 | 1.002 ± 0.01 | 0.504 ± 0.029 | 0.242 ± 0.035 | 0.023 ± 0.07 |
| | 150 | 1.004 | 1.02 | 1.02 | 1.88 ± 0.03 | 0.416 ± 0.064 | 0.243 ± 0.028 | 0.01 ± 0.008 |
| Biospal 1200S | 30 | 1.05 | 1.09 | 1.01 | 1.03 ± 0.04 | 0.46 ± 0.03 | 0.0075 ± 0.004 | 0 |
| | 150 | 1.00 | 1.02 | 1.02 | 0.855 ± 0.08 | 0.35 ± 0.07 | 0.0047 ± 0.0009 | 0 |
| Biospal 1800S | 30 | 1.04 | 1.07 | 1.06 | 1.02 ± 0.06 | 0.35 ± 0.02 | 0.05 ± 0.006 | 0 |
| | 150 | 1.01 | 1.01 | 1.02 | 0.87 ± 0.02 | 0.21 ± 0.04 | 0.01 ± 0.008 | 0 |
| CS Ultrafilter | 30 | 1.01 | 1.07 | 1.08 | 0.41 ± 0.02 | 0.015 ± 0.001 | 0 | 0 |
| | 150 | 1.06 | 1.06 | 1.06 | 0.215 ± 0.03 | 0 | 0 | 0 |
| PAN 150 | 30 | 1.08 | 1.03 | 1.02 | 0.44 ± 0.024 | 0.003 ± 0.0006 | 0 | 0 |
| | 150 | 1.02 | 1.01 | 1.03 | 0.42 ± 0.036 | 0.005 ± 0.07 | 0.009 ± 0 | 0 |

## References

1  Schütterle, G.; Wizemann, V.; Seyffart, G.: Hemodiafiltration. Proc. 1. Symp. Giessen 1981 (Hygieneplan, Oberursel 1982).
2  Fischbach, M.; Attal, Y.; Geisert, J.: Hemodiafiltration versus hemodialysis in children. A twelve-months experience. Int. J. pediat. Nephrol. *5,3:* 151−154 (1984)
3  Fischbach, M.; Hamel, G.; Geisert, J.: Hemodiafiltration: an optimal method of epuration of children? Kidney int. (submitted for publ.).

Dr. M. Fischbach, Service de Pédiatrie 3, Nephrologie et Dialyse pour Enfants, Hôpital Hautepierre, Avenue-Molière, F-67098 Strasbourg Cédex (France).

Contr. Nephrol., vol. 46, pp. 169–173 (Karger, Basel 1985)

# Performance Characteristics of the Hemoflow F 60 in High-Flux Hemodiafiltration

*B. von Albertini, J.H. Miller, P.W. Gardner, J.H. Shinaberger*

Wadsworth V.A. Medical Center and Dept. of Medicine, UCLA,
Los Angeles, Calif., USA

## Introduction

The availability of a new membrane with greater diffusive and convective permeability offers an opportunity to explore the clinically obtainable efficiency of dialysis. The technical aspects of conventional dialysis have been designed around comparatively moderate blood and dialysate flows as well as membrane surface area and hydraulic permeability. These choices, influenced in part by early observations of clinical intolerance to rapid treatments [1, 2], limit efficiency and therefore efforts to shorten treatment time. Better treatment tolerance, however, is observed with newer techniques such as bicarbonate dialysis [3], and shorter treatments with markedly increased efficiency have been shown to be well tolerated with hemofiltration [4, 5] and hemodiafiltration [6, 7].

From a technical standpoint, efficiency of solute removal in the extracorporeal circuit depends on two independent variables: the rates of blood and dialysate flows and the surface area, permeability and flow geometry of the membrane [8]. The present clinical study was undertaken to explore the extent of overall efficiency to be gained by augmenting all of the above factors.

## Methods

In order to optimally exploit membrane permeability, hemodiafiltration for simultaneous diffusive and convective solute transport was used. The study was performed in three stable male ESRD patients, who have mature forearm Cimino Brescia fistulae. Cannulation

Table 1. Whole blood clearances for urea, creatinine and phosphorus in high flux hemodiafiltration

| $Q_B$ | $Q_D$ | $Q_F$ | $Cl_{BUN}$ | $Cl_{CR}$ | $Cl_P$ |
|---|---|---|---|---|---|
| 500 | $1{,}020 \pm 15$ | $130 \pm 17$ | $431 \pm 8$ | $384 \pm 8$ | $335 \pm 35$ |
| 630 | $1{,}006 \pm 11$ | $146 \pm 18$ | $514 \pm 12$ | $431 \pm 7$ | $399 \pm 29$ |

Hemoflow F 60, $2 \times 1.25$ m$^2$ in series. $Q_B$ = Blood flow; $Q_D$ = dialysate flow; $Q_F$ = ultrafiltration rate in first device, all ml/min, n = 5; mean ± SD.

of the angio access was done with 14-gauge, 1-inch needles (Lifepath[TM], B D Drake Willock, Portland, Oreg.). Hemodiafiltration was performed in a new mode described in greater detail elsewhere [9]. Two Hemoflow F 60s (Fresenius AG, Bad Homburg, FRG) were used in a serial configuration in the extracorporeal circuit. Sterile, pyrogen-free dialysate was delivered in a counter-current mode by an automated system providing volumetric control of net ultrafiltration. By adjusting differential pressures, maximal ultrafiltration was obtained in the first, while simultaneous automatic volume replacement by backfiltration of dialysate occurred in the second device. In this configuration the entire surface area (2.5 m$^2$) of both devices was available for diffusion. Dialysate composition was: Na 140, K 2, Ca 4.5, Mg 0.7, Cl 108.2, HCO$_3$ 35, Acetate 4 mEq/l and glucose 100 mg/dl.

During the clinical treatments, the performance was routinely measured by whole blood clearances of blood and dialysate [9]. Blood flow was measured by using occlusive blood pumps calibrated at the same input and output pressures and temperatures as encountered clinically. Dialysate flow rates were measured by timing the strokes of the proportioning cylinders whose stroke volume had previously been measured. Net weight loss was determined from the slope of a continuous recording of the weight of the bed scale plus the patient. Ultrafiltration rate in the first device, $Q_F$, was determined by a ball-in-tube type flow meter (F1300, Roger Gilmont Intruments, Great Neck, N.Y,) during brief periods when the dialysate flow to this filter had been bypassed, but pressures maintained at operating levels. Pressures were automatically measured using differential transducers (140 PC series, Microswitch, Freeport, Ill.) and continuously recorded by a microcomputer (TRS 80[TM], Tandy Corp. Fort Worth, Tex.). The erythrocytes were allowed to equilibrate with plasma for at least an hour before separation.

All clearance periods were analyzed for mass balance of urea [10]. Only periods with a mass balance error of less than 10% are reported.

## Results

All treatments, where the high flows and clearances were maintained throughout, were well tolerated by the patients. The overall whole blood clearances of the two filters in series at two different blood flow rates are

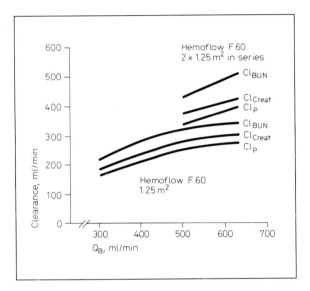

*Fig. 1.* Whole blood clearences in high flux hemodiafiltration; upper panel 2 filters in series. Lower panel: First filter in serial configuration alone. Values at $Q_B = 300$ were obtained in conventional hemodiafiltration.

summarized in table I. These results are also depicted in figure 1, where the contribution of the first device of the serial configuration, measured simultaneously, is reported separately. This allows a comparison to conventional single-device hemodiafiltration with post dilution [11]. It is apparent that at high blood flow rates the performance of the first device reaches a plateau, whereas the contribution of the second device to the overall solute transfer becomes greater.

## Discussion

The clinically obtained solute clearances in this study exceed, to our knowledge, the highest previously reported in-vivo values [4–6]. This almost threefold gain in efficiency over conventional hemodialysis was technically possible because each one of the factors determining the efficiency of extracorporeal solute removal had been augmented. Besides high blood and dialysate flows, which were obtained without difficulty, a highly permeable new membrane was used in this study. The effect of doubling surface area became apparent at high blood flow rates and suggests that the efficiency of the device in its present configuration is mainly flow dependent in the range of conventional blood flows but becomes permeability limited at higher rates. As reported elsewhere in this volume, the high hydraulic permeability of this membrane entails, as a necessary consequence of resistances to blood and dialysate flows and resulting pressure gradients, high transmembrane fluid fluxes in both directions. The described volumetrically controlled serial configuration where ultrafiltration occurs in the first and backfiltration in the second device, permits us to maximize and accurately control these high transmembrane fluxes.

The observed clinical tolerance to high solute and weight removal rates is in stark contrast to the reported experience with rapid conventional dialysis and must be related to the only two fundamentally different features in our approach: bicarbonate dialysate and a greater convective solute transport component.

## References

1    Maher, J.F.; Schreiner, G.E.: Hazards and complications of dialysis. New Engl. J. Med. *273:* 370–377 (1965).

2    Arieff, A.I.; Massry, S.G.; Barrientos, A.; Kleeman, C.R.: Brain water and electrolyte metabolism in uremia: effects of slow and rapid hemodialysis. Kidney int. *4:* 177–187 (1973).

3    Graefe, V.; Follette, W.C.; Vizzo, J.E.; Goodisman, L.D.; Scribner, B.H.: Reduction in dialysis induced morbidity and vascular instability with the use of bicarbonate in dialysis. Proc. clin. Dial. Transplant Forum *6:* 203–207 (1976).

4    Geronemus, R.; von Albertini, B.; Glabman, S.; Bosch, J.P.: High flux hemofiltration: Further reduction in treatment time. Proc. clin. Dial. Transplant Forum *9:* 125–127 (1979).

5    Shaldon, S.; Beau, M.C.; Deschodt, G.; Mion, C.: Mixed hemofiltration: 18 months experience with ultrashort treatment time. Trans. Am. Soc. artif. internal. Organs *27:* 610–612 (1981).

6    Wizemann, V.; Kramer, W.; Knopp, G.; Rawer, P.; Mueller, K.; Schütterle, G.: Ultrashort hemodiafiltration: efficiency and hemodynamic tolerance. Clin. Nephrol. *19:* 24–30 (1983).

7    Albertini, B. von; Miller J.H.; Gardner, P.W.; Shinaberger, J.H.: High-flux hemodiafiltration: under 6 hours/week Rx. Trans. Am. Soc. artif. internal. Organs *30:* (in press, 1984).

8    Michaels, A.S.: Operating parameters and performance criteria for hemodialyzers and other membrane-separation devices. Trans. Am. Soc. artif. internal. Organs *12:* 387–392 (1966).

9  Miller, J.H.; Albertini, B. von; Gardner, P.W.; Shinaberger, J.H.: Technical aspects of high-flux hemodiafiltration for adequate short (under 2 hours) treatment. Trans. Am. Soc. artif. internal. Organs *30:* (in press, 1984).

10 Henderson, L.W.: Hemodialysis: rationale and physical principles; in Brenner, Rector, The kidney, p. 1649 (Saunders, Philadelphia 1976).

11 Leber, H.W.; Wizemann, V.; Goubeaud, G.; Rawer, P.; Schütterle, G.: Hemodiafiltration: a new alternative to hemofiltration and conventional hemodialysis. Artif. Organs *2:* 150−153 (1978).

Dr. B. von Albertini, Nephrology Section W 111L, Wadsworth V.A. Medical Center, Wilshire and Sawtelle Blvds., Los Angeles, CA 90073 (USA)

Contr. Nephrol., vol. 46, pp. 174–183 (Karger, Basel 1985)

# Investigation of the Permeability of Highly Permeable Polysulfone Membranes for Pyrogens

*H. Klinkmann, D. Falkenhagen, B.P. Smollich*

Department of Nephrology, Wilhelm-Pieck-University, Rostock, GDR

## Introduction

The treatment of acute and chronic renal failure is primarily based upon procedures that employ diffusive as well as convective transport processes. For the detoxification of uremic patients, hemodialysis (HD), hemofiltration (HF) as well as hemodiafiltration (HDF) and the different forms of peritoneal dialysis (PD) are used.

In these various mechanical methods of detoxification the blood is indirectly or directly brought into contact with solutions that either remove waste products (HD, PD) or reconstitute the original flow properties of the blood, refilling the convectively diminished blood volume (HF, HDF).

The preparation and processing of these solutions has the prevailing risk that the solutions will be contaminated with pyrogens, i.e. substances which, upon entering the human body, can produce different toxic effects, such as fever, chills, nausea or anaphylactic reactions. Therefore, it is the undertaking of physicians and technicians to reduce the amount of potential pyrogens in the aforementioned solutions as completely as possible so that the risk of these reactions can be eliminated. Of course, a complete elimination of pyrogenic substaces is of great importance in the production of parenteral solutions.

The testing for pyrogens therefore belongs to the standard of all known pharmacopoeias. The USP XII, valid from November 1942 [1], was the first pharmacopeia in the world to cover testing requirements for pyrogens.

As membranes for HF and high-flux HD became available, the interest in using these membranes also for the elimination of pyrogens was ever increasing. Possibilities for use of these uncomplicated filters could include the preparation of substitution fluids for HF and HDF, solutions for PD and of solutions or drugs for parenteral applications.

The purpose of this study was to test a newly developed polysulfone membrane for its permeability for pyrogens. The polysulfone membrane evaluated is used in the high-flux dialyzers F 40 and F 60 (Fresenius AG, Oberursel, FRG) which have been successfully applied clinically in high-flux HD and HDF [2, 3].

### Occurrence, Structure and Mode of Action of Pyrogens

In most cases, the existence of pyrogens is associated with the presence of bacteria. Aside from these so-called bacterial pyrogens there are a number of other pyrogens that are not of bacterial origin. In general, pyrogens can be classified in exogenous and endogenous pyrogens, whereby exogenous pyrogens include chemical pyrogens (e.g. LSD), pyrogens from virus, bacteria, mycobacteria, fungi and plants (e.g. alkaloids). The exogenous pyrogens are considered to be the contaminating agents that are able to cause different reactions in the organism.

Pyrogens from gram-negative bacteria are the most important form. They constitute the sub-group that up to now has been the most thoroughly studied. They are almost exclusively responsible for the occurence of undesirable side effects after parenteral application of contaminated infusion or injection solutions and therefore deserve special attention in medical and pharmaceutical practice.

These pyrogens are protein-lipid-lipopolysaccharide complexes (fig. 1) of high molecular weight. The lipopolysaccharid with its various components is responsible for the immunological as well as the endotoxic properties of the pyrogen. It is important to realise that lipoid A, with a molecular weight of about 2,000 daltons, can be split from this big complex by acid hydrolysis. Lipoid A has pyrogenic properties, too, but the test methods (LAL-test) are sensitive for these molecules as well.

There is a time lapse between the uptake of lipopolysaccharides from gram-negative bacteria and the onset of clinical reactions. Figure 2 shows the possible pathway, by which contamination with exogenous pyrogens results in a febrile reaction [5].

It should be mentioned that the onset of fever is mediated by endogenous pyrogens and their effect on thermosensitive neurones in the hypothalamic region. These endogenous pyrogens have recently been held responsible for some side effects of HD therapy which can be observed 1 to 3 hrs after the beginning of treatment. It has been hypothesized that Interleucin 1, produced by monocytes, plays an important role in this process [6].

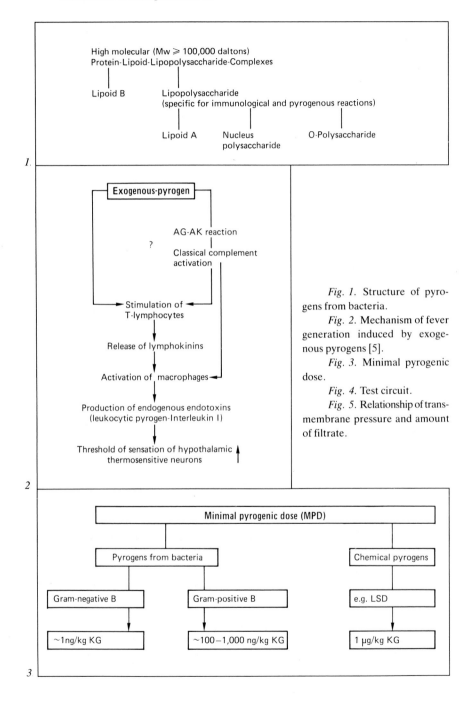

*Fig. 1.* Structure of pyrogens from bacteria.

*Fig. 2.* Mechanism of fever generation induced by exogenous pyrogens [5].

*Fig. 3.* Minimal pyrogenic dose.

*Fig. 4.* Test circuit.

*Fig. 5.* Relationship of transmembrane pressure and amount of filtrate.

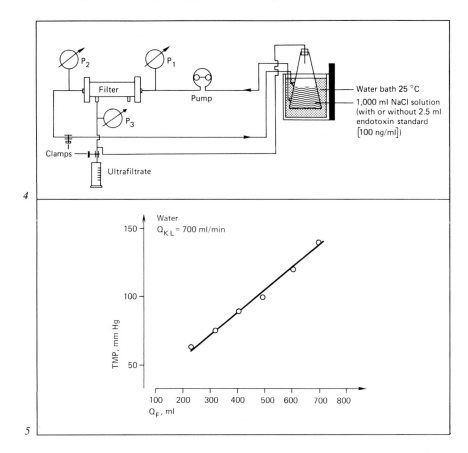

4

5

It is interesting and important to keep in mind the so-called minimal pyrogenic dose that leads to a rise in body temperature. This amount, as determined in animal tests (rabbits), has been consistently stated throughout literature as 1.0 mg/kg body weight [7, 8]. The sensitivity towards endotoxins from gram-positive bacteria or towards chemical pyrogens is much lower (fig. 3).

### Material and Methods

Lipopolysaccharide *E. coli* 026:B 6 (Sigma, St. Louis, Mo.) was used to study the permeability for endotoxins. We prepared an endotoxin-standard that allowed the production of a final solution with concentration of 100 ng/ml by adding 2.5 ml of the standard to 1,000 ml of pyrogen-free saline.

*Table I.* Permeation of pyrogens in the MLW 1.3 (n = 11)

| TMP, mm HG | $Q_{KL}$, ml/min | LAL test | | | |
|---|---|---|---|---|---|
| | | 10 min | 20 min | 60 min | 120 min |
| 50 | 500 | – | – | – | – |
| 100 | 500 | – | – | – | – |

*Table II.* Permeation of pyrogens in the F 60 (n = 5)

| | TMP, mm Hg | $Q_{KL}$, ml/min | LAL test (n = 3) | | | |
|---|---|---|---|---|---|---|
| | | | 10 min | 30 min | 60 min | 120 min |
| 1st filter | 50 | 500 | – | – | – | – |
| | 75 | 500 | + | + | + | + |
| | 100 | 500 | + | + | + | + |
| 2nd to 5th filters | 50 | 500 | + | + | + | + |
| | 75 | 500 | + | + | + | + |
| | 100 | 500 | + | + | + | + |

The test circuit is shown schematically in figure 4.

The filters used were (1) Dialyzer MLW 1.3 (VEB MLW Keradenta-Werk, Radeberg, GDR) with a surface area of 1.3 m and hollow fibers from regenerated cellulose. (2) Hemodiafilter F 60 (Fresenius AG, Bad Homburg, FRG) with an effective surface area of 1.25 m² and polysulfone hollow fibers.

An LAL test (Limulus Amebocyte Lysate) (Pyroquant GmbH, Walldorf, FRG) was applied to detect the presence of endotoxins in the circuit and in the filter. For the preparation of the standards and for diluting the LAL test as well as for the control experiments without endotoxins, pyrogen-free water and pyrogen-free isotonic saline solutions were used.

The experimental conditions can be seen in figure 4. Temperature was kept constant at 25 °C. 1,000 ml of pyrogen-free or endotoxin containing solution were perfused by means of a roller pump (Fresenius AG, Bad Homburg, FRG) through the filter to be tested. Sterile tubing sets for dialysis were used for the experiment.

Transmembrane pressure (TMP) was varied from 50 to 100 mm Hg and the flow rate in the circuit ($Q_{KL}$) from 200 to 700 ml/min (fig. 5).

Samples were taken from the filtrate after 10, 30, 60 and 120 min. In some cases, the experiments were carried out for over 6 days. The F 60 filters tested were filled with formaline and consequently retested for their permeability for pyrogen.

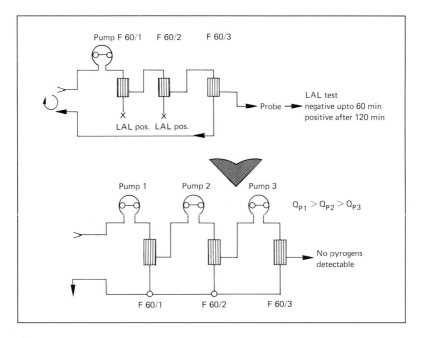

*Fig. 6.* F 60 test circut arranged in series.

## Results

(1) Testing of the blood tubing sets (n = 3) without filter. No pyrogenicity of the blood tubing sets could be detected.

(2) Testing of the dialyzer MLW 1.3 (n = 11). (a) Testing without endotoxin: The LAL-test did not respond, i.e. contact with and filtration of pyrogen-free water through the dialyzer MLW 1.3 did not result in a positive LAL test. (b) Permeation of endotoxin: There were no indications of a permeation of endotoxins through the cellulosic membrane of the MLW 1.3. Table I illustrates this result for a $Q_{KL}$ of 500 ml/min.

(3) Testing of the hemodiafilter F 60 (n = 5). (a) Determination of filtration rate as a function of TMP: The experiments were repeated at different values for $Q_{KL}$ (200, 400, 500, 700 ml/min). Even at a $Q_{KL}$ of 700 ml/min and a closed clamp at the exit of the filter, a TMP of only 140 mm Hg could be achieved. Figure 5 shows filtration rate for the F 60 at a $Q_{KL}$ of 700 ml/min as a function of TMP. (b) Control test: The LAL-test did not react with fluid from the circuit or from the filtrate. (c) Determination of

permeability for pyrogens: Table II shows the results at a $Q_{KL}$ of 500 ml/min. At a TMP of 50 mm Hg, the LAL test did not detect pyrogenic substances in the filtrate in the concentration given by the sensitivity limits. An increase of TMP then resulted in a permeation of pyrogenic substances. All other filters examined showed a permeability for pyrogens under the applied operating conditions. (d) Determination of the sieving coefficient: The sieving coefficient for the endotoxin used was determined for 2 filters. The concentration was derived by diluting the sample until the LAL test would no longer detect any pyrogenic substances and by multiplying the dilution factor with the limit of sensitivity of 0.006 ng/ml. Concentrations in the filtrate were estimated to be 0.072 ng/ml and 0.192 ng/ml, corresponding to a sieving coefficient of 0.001 to 0.002. (e) Neither the filling with formaline nor a long-term use for more than 6 days changed the filter's permeability for pyrogens. (f) When 3 F 60 filters were arranged in series (fig. 6, above), the LAL test would not react with the filtrate taken from the 3rd filter after 10, 30 and 60 min, at an endotoxin-concentration in the reservoir of 100 ng/ml. The LAL test reacted, however, after 120 min. The filtrates from the 1st and the 2nd filter contained detectable pyrogenic substances from the beginning.

## Discussion

As the results for the MLW 1.3 cellulose dialyzer clearly demonstrate, the control experiments did not result in a positive LAL test.

Investigation with aqueous extracts from cellulose dialyzers showed that positive reactions of the LAL test are possible [9]. For this reason, blind tests are mandatory when membrane filters are tested for the permeability for pyrogens using the LAL test. The cellulose membrane, also, did not allow for any detectable permeation of endotoxins, with the limit of sensitivity of the pyrogen test being 0.006 ng/ml.

These results agree with data reported by other authors who tested the pyrogen permeabilities of conventional dialyzers, especially under in vivo conditions [10].

It was found consistently that even with pyrogen concentrations in the dialysate far above 100 ng/ml, only very small amounts of endotoxins can permeate the membrane. The main reason for this is the small pore-size of the cellulosic membrane used in relation to the high molecular weight of the endotoxins.

The examination of the hemodiafilter F 60 clearly demonstrated the permeability of the polysulfone membrane investigated for endotoxins.

As was mentioned previously, bacterial pyrogens can disintegrate into fragments with a molecular weight of below 10,000 daltons. In view of this, the question has to be raised, how a membrane should be able to combine, on the one hand, a complete impermeability to all sorts of pyrogens, even in the lower molecular weight, and, on the other hand, a decent permeability for higher molecular substances (molecular weight approx. 20,000 daltons). Certain higher molecular substances are increasingly discussed as possible uremic toxins [11].

Our experimental setup was aimed at simulating an on-line production of infusion fluid. It cannot be compared to the conventional operating conditions of hemodialysis – flow rates, even at a TMP of only 50 mm Hg would correspond to a filtration of about 200 ml/min from the dialysate into the blood.

Under the working conditions of HD, the amount of endotoxin that could be transported into the patients blood if a filtration from the dialysate into the blood occurred, is extremely small, especially when the low sieving coefficients of 0.001 to 0.002 are taken into consideration. The exact amount, of course, depends on the concentration of pyrogen in the dialysate. In view of the fact that at present reverse osmosis units are used almost exclusively for water preparation in dialysis centers, high concentrations of pyrogens in the dialysate are very improbable. As our own measurements disclosed, the endotoxin concentration in our reverse osmosis unit rose to only 0.03 ng/ml after disinfection was omitted for several weeks. Even if one takes the hypothesized possibility of a so-called on-line HDF with in the F 60 [12] into account, it is improbable that the minimum pyrogen dose could be reached. In addition to this, plasma proteins are adsorbed to the membrane in the in vivo situation which further diminish the sieving coefficients for pyrogens.

An exact examination of the questions raised still needs to be done. The results of the clinical applications of the F 60, reported so far, have not disclosed any pyrogen reaction [2, 3].

The results derived with filters switched in series show the possibility of eliminating the pyrogens using three F 60 hemodiafilters. The fact that a positive LAL test was obtained after 120 min in the filtrate of the third filter can be explained by the low sieving coefficients of the filters, resulting in a higher concentration of endotoxins at the membrane. Due to this, a significant amount of endotoxins is transported through the membrane, although the sieving coefficient is actually very low. Each consecutive filter was subjected to this break-through at a later time.

In order to avoid this, 3 filters and 3 pumps could be used (fig. 6, below). As part of the fluid entering the filter leaves it in the same compartment, an enrichment of endotoxins above a certain degree does not take place. The high hydraulic permeability of the polysulfone membrane should make it feasible to operate with high flow rates ($\geq$ 1 liter/min). It would be necessary to use the pumps at decreasing velocity from the first to the last filter.

A further simplification of such an installation using the F 60 can probably only be achieved by slightly diminishing the average pore size of the polysulfone membrane.

We therefore conclude that in addition to its application in HD and HF, the polysulfone membrane holds enormous interest, for instance to the production of infusion solutions in the pharmaceutical industry.


*Addendum*

In order to elucidate the clinical relevance of our findings, we have, since the presentation of our first results in May 1984, studied the permeability of the polysulfone membrane for pyrogens under boundary conditions closer to those encountered in hemodialysis. As these experiments were completed after the proceedings of the symposium went into print, they are summarized briefly here with the kind permission of the editors.

The experimental set-up is described in Figure 4. Endotoxin of *E. coli* (0.55:B5) was used at a concentration of 100 ng/ml. 'Blood' flow rate was set at 200 ml/min, trans-membrane flux at 100 ml/min. This is still a much higher value than the eventual rate of backfiltration, which, under the most unfavourable conditions, has been hypothesized to be between 10 and 20 ml/min.

In controlled experiments we could not detect the passage of any substances with a positive reaction in the LAL-test across the membrane. These results seem to indicate that, as expected from our own practical experience using the F 60 hemodialyzer, the device investigated is as safe for use in hemodialysis as other hemodialyzers equipped with conventional membranes.

One possible explanation for these results which differ from our original findings, of course, are the altered boundary conditions. This demonstrates the importance of using standardized conditions for these investigations in order to achieve comparable results.

## References

1    The Pharmacopeia of the United States of America. Twelfth Revision (USP XII). Official from November 1. 1942, Easton, Pa.
2    Falkenhagen, D.; Ahrenholz, P.; Falkenhagen, U.; Holtz, M.; Behm, E.; Roy, T.; Klinkmann, H.: Efficiency and blood compatibility of a new polysulfone high-flux-dialyzer. Artif. Organs 7: 44.
3    Streicher, E.; Schneider, H.: Stofftransport bei Hämodiafiltration. Nieren-Hochdruck-Krankh. 12: 339-342 (1983).
4    Port, F.K.; Bernick, J.J.: Pyrogen and endotoxin reactions during hemodialysis. Contr. Nephrol., vol. 36, pp. 100−106 (Karger, Basel 1983).
5    Dinarello, C.A.: Pathogenesis of fever during hemodialysis. Contr. Nephrol., vol. 36 (Karger, Basel 1983).
6    Henderson, L.W.; Koch, K.M.; Dinarello, C.A.; Shaldon, S.: Hemodialysis hypotension. The interleukin hypothesis. Blood Purification 1: 3−8 (1983).
7    Pearson, F.C., III; Weary, M.E.; Bobon, J.; Dabbah, R.: Relative potency of 'environmental' endotoxin as measured by the limulus amebozyte lysate test and the USP rabbit pyrogen test; in Endotoxins and their detection with the limulus amebocyte lysate test, pp. 65−77 (Liss, New York 1982).
8    Dressel, H.: Beiträge zur Methodik der Prüfung pyrogener Verunreinigungen unter besonderer Berücksichtigung des als Pyrogen-Arbeitsstandard verwendeten Natrium-Nukleinats B-Promotionsarbeit, Humboldt-University, Berlin.
9    Henne, W.; Schulze, H.; Pelger, M.; Tretal, J.; Sengbusch, G. v.: Hollow fiber dialyzers and their pyrogenicity testing by LAL. Symp. Hypersensitivity in Hemodialysis, Louisville 1983.
10   Bernick, J.J.; Port, F.K.; Favero, M.S.; Brown, D.G.: Bacterial and endotoxin permeability of hemodialysis membranes. Kidney int. 16: 491−496 (1979).
11   Wizemann, V.; Velcovsky, H.G.; Bleyl, H.; Brüning, S.; Schütterle, G.: Removal of hormones by hemofiltration and hemodialysis with a highly permeable polysulfone membrane. (this issue).
12   Schmidt, M.; Baldamus, C.A.; Schoeppe, W.: Characterization of solute and solvent kinetics in hemodialyzers with highly permeable membranes. Am. Soc. artif. internal Organs, Abstr. 13: 55 (1984).

OMR Prof. Dr. sc. med. H. Klinkmann, Klinik für Innere Medizin der
Wilhelm-Pieck-Universität Rostock, Ernst-Heydemann-Strasse 6,
DDR-2500 Rostock (GDR)

Contr. Nephrol., vol. 46, pp. 184–186 (Karger, Basel 1985)

# Clinical Evaluation of a Multipurpose Dialysis System Adequate for Hemodialysis or for Postdilution Hemofiltration/Hemodiafiltration with On-Line Preparation of Substitution Fluid from Dialysate

*B. Canaud[a], Q.V. N'Guyen[a], C. Lagarde[a], F. Stec[a], H.D. Polaschegg[b], C. Mion[a]*

[a]Service de Nephrologie, Hôpital Lapeyronie, Montpellier, France;
[b]Fresenius AG, Bad Homburg, FRG

This is the preliminary clinical study of a new hemodialysis/hemofiltration/hemodiafiltration system based on the Fresenius dialysis machine A2008 C.

One of the main drawbacks of hemodiafiltration (HDF) is its higher cost caused mainly by the cost of large amounts of industrially produces sterile and pyrogen-free substitution fluid to compensate for the high ultra-filtration rate. One-line preparation of substitution fluid directly from the dialyzing fluid by filtration offers therefore an attractive alternative.

The equipment used was a Fresenius dialysis machine A2008 C adapted with a sterile filter assembly as shown in figure 1. The adaption consists of a substitution pump module similar to the blood pump module of the machine, two F 60 hemodiafilters (Fresenius), a final sterile filter (0.2 µm) and a valve block which allows the mode of operation of the machine to be changed from HD to HDF or HF. Due to the volumetric fluid-balancing system of the dialysis machine used, the amount of fluid pumped by the substitution pump into the venous drip chamber is ultrafil-trated automatically and simultaneously via the two F 60, leaving the blood compartment of the second hemodiafilter used in the extra circuit. Vol-umetrically mixed bicarbonate dialysate was used for dialysis as well as for substitution. Water treatment consisted of a softener, activated charcoal, a

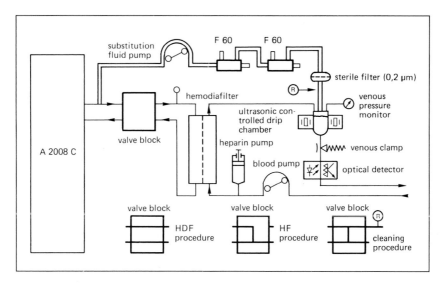

*Fig. 1.* This figure shows the three main components of the multipurpose dialysis system: 1. The valve block permitting the instantaneous change of the operating mode (HD to HDF to HF). 2. The A 2008 C insuring the volumetric (inlet and outlet) fluid balance through the hemodiafilter. 3. The substitution circuit with 2 F 60 and 0.22 μm filter in series, allowing absolute filtration and bacteriological monitoring of the infusate fluid.

1-μm filter and reverse osmosis system including bacteriological filtration with 0.45 and 0.22 μm filters.

The serial connection of the two hemodiafilters and an additional 0.22-μm sterile filter offered optimal safety even in the case of a first failure, e.g. a rupture of the membrane of one filter. The F 60 filters were changed every 15 days, the final 0.22-μm sterile filter was changed prior to each treatment. After some preliminary tests to find the optimal disinfection procedure for the system, we decided to use the following procedure: rinse with peracetic acid (Puristeril®, Fresenius) after each treatment, followed by 85 °C pasteurization for 20 min and 85 °C pasteurization for 20 min in the evening followed by a 2.5% formalin exposure overnight.

The clinical study was carried out on 4 hemodialysis patients: 1 male and 3 females, mean age 53 years. The patients had been on maintenance dialysis for an average period of 46.5 months, ranging from 16 to 74. All patients had a blood connection with arteriovenous fistulas.

The treatment protocol consisted of three weekly, 4 hour HDF sessions, the blood flow rate ($Q_B$) was maintained at 400 ml/min and the filtration rate at 69 ml/min, permitting an exchange volume of 16 liters per session. Altogether 66 HDF sessions were conducted in 4 patients within a 5 week period.

The average clearances were as follows: urea 317 ml/min which is equal to a 76% overall extraction ratio, and creatinine 292 ml/min which is equal to a 70% overall extraction ration.

## Conclusions

Clinical tolerance was excellent, high clearance rates were optained and no febrile reactions were observed. No ruptures of the membranes of the repeatedly used hemodiafilters were observed. In our opinion the disinfection procedure of the sterile filter circuit needs further optimization.

Dr. B. Canaud, Service de Nephrologie, Hôpital Lapeyronie,
F-34000 Montpellier (France)

# Subject Index